NETWORK PROGRAMMIG

By

Dr. Sachin Kumar
[M.Tech(CS), Ph.D(CS)]
Associate Professor
***RKGIT,** Ghaziabad*

Kadambri Agarwal
M.Tech. (CSE)
Associate Professor
***RKGITW,** Ghaziabad*

S.K. KATARIA & SONS®
Publisher of Engineering Books
4885/109, Prakash Mahal, Dr. Subhash Bhargav Lane,
Daryaganj, New Delhi-110002 (India)
Phone: +91-11-43551243; Mobile: +91-9871775858, 7678425209
e-mail: katariabook@gmail.com, katariabooks@yahoo.com
Website : www.skkatariaandsons.com
(e-books, subscriptions and bulk buying also available on our website)

Published by :
S.K. Kataria & Sons®
4885/109, Prakash Mahal, Dr. Subhash Bhargav Lane,
Opposite Delhi Medical Association, Daryaganj, New Delhi-110002
Phone: +91-11-43551243; Mobile: +91-9871775858, 7678425209
e-mail: katariabook@gmail.com, katariabooks@yahoo.com

First Edition : 2011, Reprint : 2026

Price: ₹ 495/-

Laser Typesetting by : Arpit Printographers

PREFACE

Network Programming is one of the interesting subject in B.Tech. curriculum. In this subject, focus is on the Basic Concepts of Network, Internetworking, IP Addresses and Implementation on Unix and Linux.

This book is written as a textbook for B.Tech (VIth) semester students of M.D. University, Rohtak and a reference book for other universities in which this subject is a part of curriculum.

This book has been written in a very simple and lucid manner. Every effort has been made to make the treatments simple and comprehensive, throughout the book, the stress has been given on fundamental concepts through illustrative examples.

This book has been divided into Nine Chapters.

- In each chapter, sufficient worked examples have been given. At the end of each chapter few very important solved problems and unsolved exercise have been given.
- Highlights at the end of each chapter has been specially written for the students so that they can grasp the summary of each chapter, Students are advised to go through highlights.
- Appendix at the end provides additional concepts. The book provides numerous examples.

We will feel highly obliged to all readers for their constructive suggestions healthy criticism of the book which will go a longway in the improvement of text. Students are advised to read solved questions and highlights at the end of chapters carefully.

Previous year question papers with cited solution are provided at the end. Students are advised to go through them.

AUTHORS

ACKNOWLEDGEMENT

We take this opportunity to express our gratitude and thanks to the respected Dr. A.K. Sharma, Prof. and Head YMCA, Faridabad for his valuable technical suggestions and constant encouragement, without which this book would not have come into existence.

We are specially grateful to Dr. B.K. Gupta, Advisor, RKG group, Dr. Laxman Prasad, Director, RKGIT, Dr. Neet Awasthy, Director RKGITW for constant inspiration and encouragement.

Special thanks to Dr. VSP Srivastav, Head, Computer Division, IGNOU, Dr. M.K. Gupta, Professor CCS University, Dr. D.K. Nobiyal, Associate Prof., JNU, Dr. Vipin Kumar (H.O.D. Physics KIET), Prof. Sanjay Goel (KIET), Prof. Gopindra Kumar (RKGIT), for their guidance and valuable suggestions.

We deeply express my heartful thanks to the publishers S.K. Kataria & Sons for publishing this book in such a beautiful get up and well in time.

Lastly, we are also greatful to Mr. Sachin and his staff for their hard work.

AUTHORS

CONTENTS

UNIT-I

1. INTRODUCTION TO NETWORKING
2. TCP/IP PROTOCOL ARCHITECTURE
3. IP DATAGRAM FORMAT

Introduction to Networking

Chapter 1

INSIDE THIS CHAPTER

1.1. INTRODUCTION **[MDU Dec. 2008; May 2006]**

A computer network consists of a collection of computers, printers and other equipment that is connected together so that they can communicate with each other. Figure 1.1 gives an example of a network in a school comprising of a local area network or LAN connecting computers with each other, the internet, and various servers.

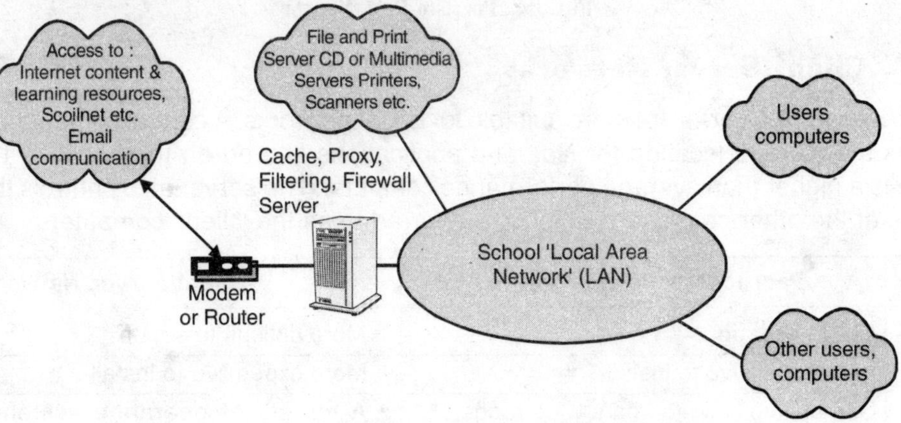

Fig. 1.1. Network.

Broadly speaking, there are two types of network configuration

- ➢ Peer-to-peer networks and
- ➢ Client/server networks

1.1.1. Peer-to-Peer Networks

Peer-to-peer networks are more commonly implemented where less then ten computers are involved and where strict security is not necessary. All computers have the same status, hence the term 'peer', and they communicate with each other on an equal footing. Files, such as word processing or spreadsheet documents, can be shared across the network and all the computers on the network can share devices, such as printers or scanners, which are connected to any one computer.

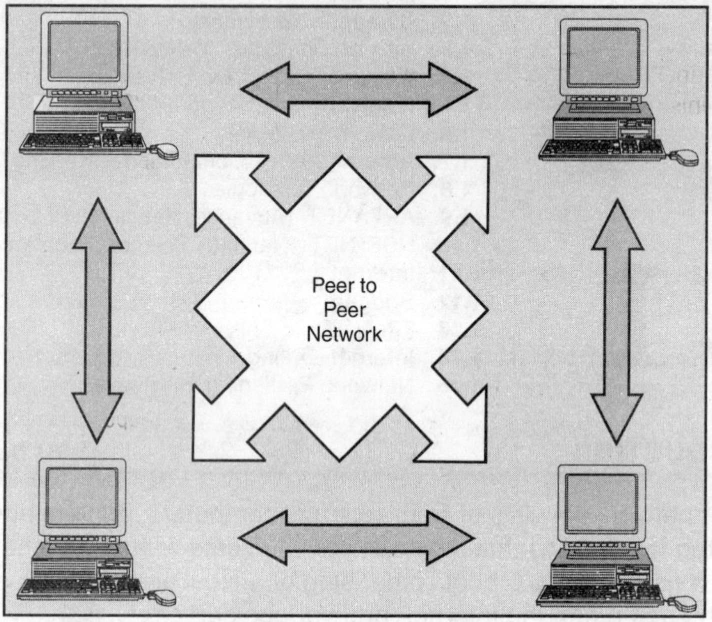

Fig. 1.2. Peert-to-Peer Networks.

1.1.2. Client/Server Networks

Client/Server Networks are more suitable for larger networks. A central computer, or 'server', acts as the storage location for files and applications shared on the network. Usually the server is a higher than average performance computer. The server also controls the network access of the other computers which are referred to as the 'client' computers.

S. No.	Peer-to-Peer Networks	Client/Server Networks
1.	Easy to set up	More difficult to set up
2.	Less expensive to install	More expensive to install
3.	Can be implemented on a wide range of operating systems	A variety of operating systems can be supported on the client computers, but the server needs to run an operating system that supports networking.
4.	More time consuming to maintain the software being used (as computers must be managed individually.	Less time consuming to maintain the software being used (as most of the maintenance is managed from the server).

S. No.	Peer-to-Peer Networks	Client/Server Networks
5.	Very low levels of security supported or none at all. These can be very cumbersome to set up, depending on the operating system being used.	High levels of security are supported, all of which are controlled from the server. Such measures prevent the deletion of essential system files or the changing of settings.
6.	Ideal for networks with less than 10 computers.	No limit to the number of computers that can be supported by the network.
7.	Does not require a server.	Requires a server running a server operating system.
8.	Demands a moderate level of skill to administer the network.	Demands that the network administrator has a high level of IT skills with a good working knowledge of a server operating system

1.2. COMPUTER NETWORKING [MDU Dec. 2008; Dec. 2006]

Interconnection of autonomous computer is called computer networking. Interconnection means information can be enchanged from one machine to other and autonomous means one computer cannot start or stop the other.

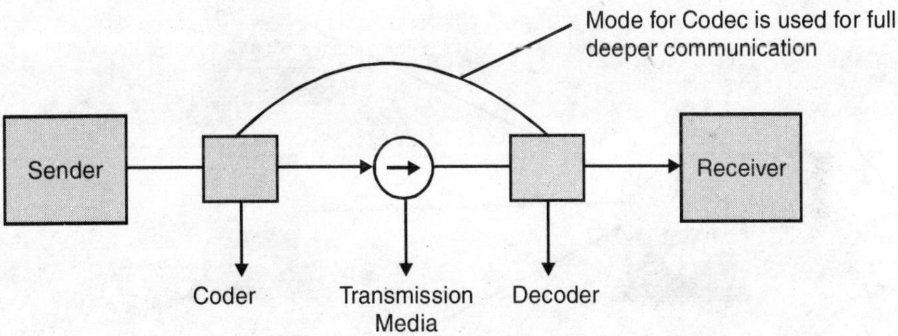

Fig. 1.3. Computer Network.

1.2.1. Difference Between Distributed System and Computer Network

In distributed system, the existence of multiple autonomous computers is not visible. It looks like a virtual uniprocessor. Here *Implicity* everything is done on a best machine. It is up to the Operating System to select the best processors. In affect, adistributed system is software system built on top at a network. The software gives it a high degree of *cohesiveness* and transparency.

In network system, everything is implicitly done by the user. User decide the best processor to be used for file transfer.

1.3. USE OF COMPUTER NETWORKING

(i) Reliability
(ii) Saving money
(iii) Resource sharing

1.4. TYPES OF COMPUTER NETWORKING

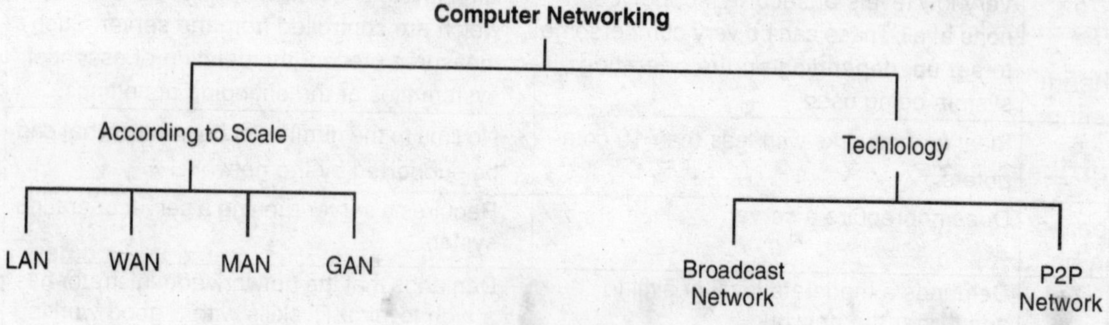

Fig. 1.4. Types of Networking

1.4.1. Broadcast Network

Broadcast network have a single communication channel that is shared by all the machines on the network.

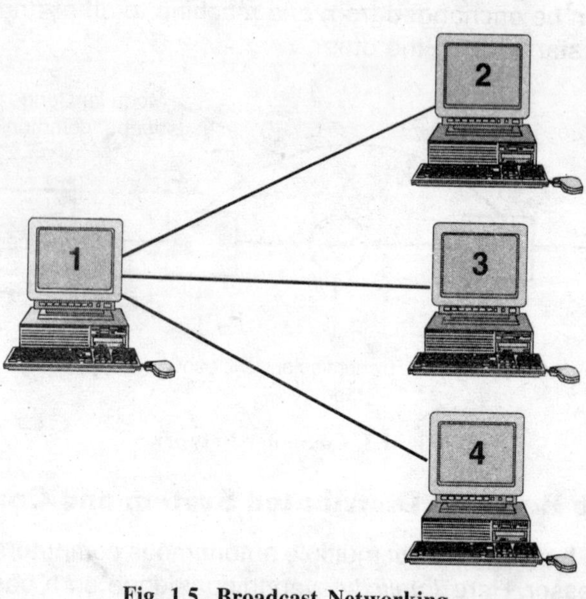

Fig. 1.5. Broadcast Networking.

Short message called **packet** are sent by any machines are received by all the other. A packet consist of two parts **header** and **data.**

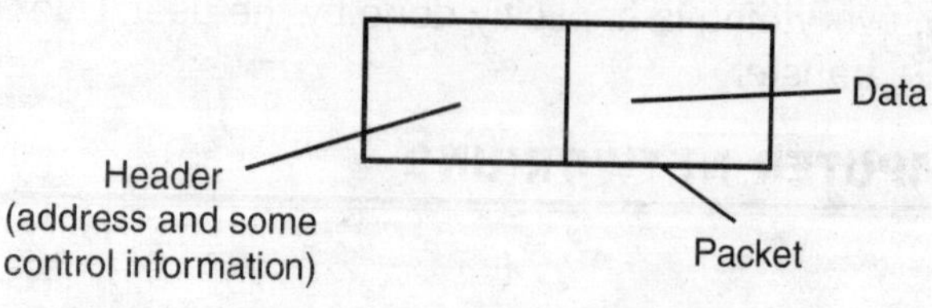

Fig. 1.6. Packet Networking.

Upon receiving a packet, machine checks the address field, if the packet is intended for itself, it processes the packet, other wise igonored it.

Broadcasting

When a packet is received and processed by every machine OL network is called broadcasting.

Multicasting

If packets are processed by subset of network then it is multicasting. Broadcast network can be further divided into static and dynamic, depending on how the channel is allocated

Fig. 1.7.

Static

Time is fined for each machine although it may be different for each machine.

Dynamic

Time on demand (fined). Round RO6 in is used. Further dynamic allocation methods for a common channel are either centralized or decentralized.

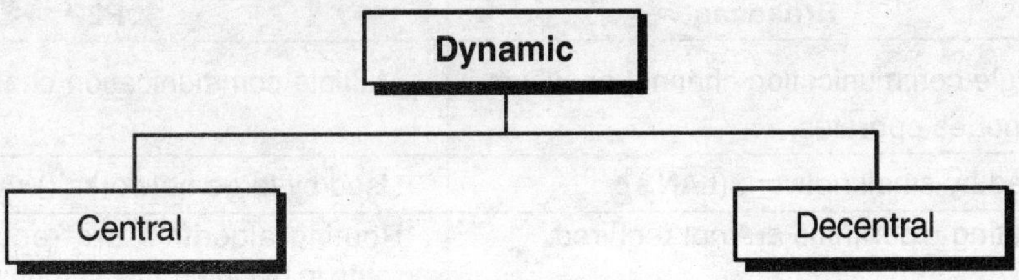

Fig. 1.8.

Centralised

There is only single machine which decides transmission of data by next machine. The machine itself does not transfer any data.

Dencentralized

Each and every machine can take the decision for transmitting the data.

- Broadcast network technology is generally used by small size network LAN's various topologies are possible for barodcast LANS. Two of them are tokenbus (802.4) and (802.5) Ring. No need of routing algorithm is broadcast network so network layer is after thin or absent. MAC sublayer is used to solve channel all can problem.
- P2P is network is used by WAN's.

1.5. POINT TO POINT NETWORK

Point to point network consists of many connection between individual pair of machine many large geographically localised network (WAN) tend to use P2P.

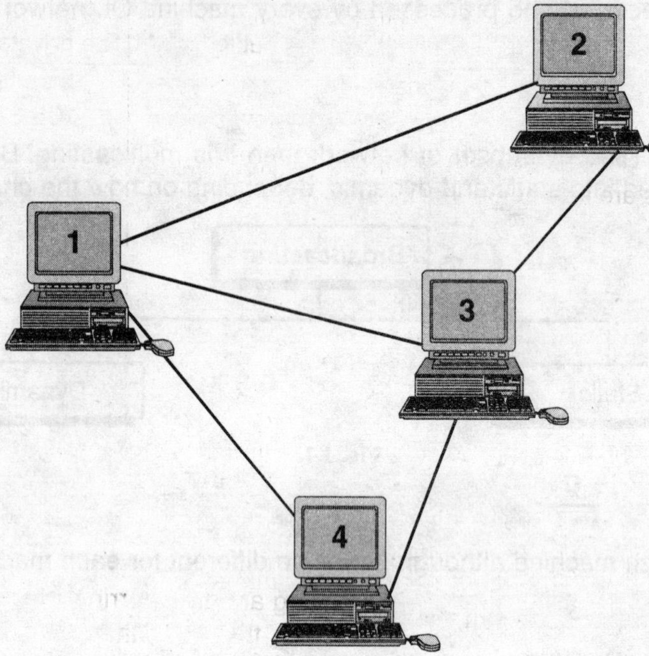

Fig. 1.9.

S. No.	Broadcast	P2P
1.	Single communication channel on which all nodes operates.	Multiple communication channel.
2.	Used by small network (LAN's).	Used by large networks (WAN's).
3.	Routing algorithms are not required.	Routing algorithm are required to decide route to decide route as multiple connections are there.
4.	Network layer is often thin or absent.	Network layer is present.
5.	MAC sublayer for channel allocation problem.	Routing is done by network layer.

1.6. LAN, WAN, MAN

LAN (Local Area Network), WAN (Wide Area Network) and MAN (Metropolitan Area Network) are different in their

(a) Size

(b) Transmission technology

(c) Topology

1.6.1. Difference between LAN, WAN, MAN

S. No.	LAN	WAN	MAN
1.	Privately owned networks within a building or office up to few km in size.	WAN spans a large geographical area often a country or continent. It is public.	Bigger version LAN. It covers a group of offices or city. It can be private or publi.
2.	LANs often uses a transmission technology consisting of single cable to which all machines are attached.	It consist of Host and Subnet further subnet is collection of routers and transmission lines.	It uses a standard 802.6 (DQDB) distributed queue dual bus. DQDB consists of two unidirectional buses (cables) to which all computers are connected.
3.	Broad cost LAN's has two important topologies 802.4 and 802.5.	Possible topologies for WAN using P2P are start, ring, tree, complete etc.	
4.	Used broadcast.	Uses P2P.	May use P2P or broadcast.

1.7. ADVANTAGES OF COMPUTER NETWORKING [MDU Dec. 2008; Dec. 2006]

1. Speed

Networks provide a very rapid method for sharing and transferring files. Without a network, files are shared by copying them to floppy disks, then carrying or sending the disks from one computer to another. This method of transferring files in this manner is very time-consuming.

2. Cost

The network version of most software programs are available at considerable savings when compared to buying individually licensed copies. Besides monetary savings, sharing a program on a network allows for easier upgrading of the program. The changes have to be done only once, on the file server, instead of on all the individual workstations.

3. Centralized Software Management

One of the greatest benefits of installing a network at a school is the fact that all of the software can be loaded on one computer (the file server). This eliminates that need to spend time and energy installing updates and tracking files on independent computers throughout the building.

4. Resource Sharing

Sharing resources is another area in which a network exceeds stand-alone computers. Most schools cannot afford enough laser printers, fax machines, modems, scanners, and CD-ROM players for each computer. However, if these or similar peripherals are added to a network, they can be shared by many users.

5. Flexible Access

School networks allow students to access their files from computers throughout the school. Students can begin an assignment in their classroom, save part of it on a public access area of the network, then go to the media center after school to finish their work. Students can also work cooperatively through the network.

6. Security

Files and programs on a network can be designated as "copy inhibit," so that you do not have to worry about illegal copying of programs. Also, passwords can be established for specific directories to restrict access to authorized users.

1.7.1. Disadvantages of Computer Networking [MDU May 2008]

1. **Security Issue :** Computer Network is a threat for security of data as any unauthorized user can also access some important data.
2. **Verus Threat :** If one computer in a network get infected by virus there is possibility of virus infecting all the computers in network.
3. **Network Halt :** Any fault in connecting devices like switches, hubs etc. and lead to a whole network halt.
4. **Expensive Setup :** The initial setup cost of a computer network can be high depending on the number of computers to be connected.
5. **Dependency on Main File Server :** In case the main file server of computer network breaks down, the sysetm becomes useless. In case of big networks, the file server should be a powerful computer.

1.8. NETWORK TOPOLOGIES

Topology literally means the shape or architecture that is used to joining the network.

OR

Topology is way in which the end points or a stations of a network are linked together.

1. Star Topology

In a star configuration of computer network, there is a host computer which is attached to local is a host computer which is attached to local computer through multiple communication lines local computers are not linked directly to each other and can communicate via Lostcomp (lub/switch). Rooting function is performed by lost computer

Advantages

(a) Star topology has minimal line cost because only n – 1 lines are required for connecting n-nodes.

(b) It any local computer fails, the remaining portion of network in unaffected.

(c) Addition of new node is not difficult.

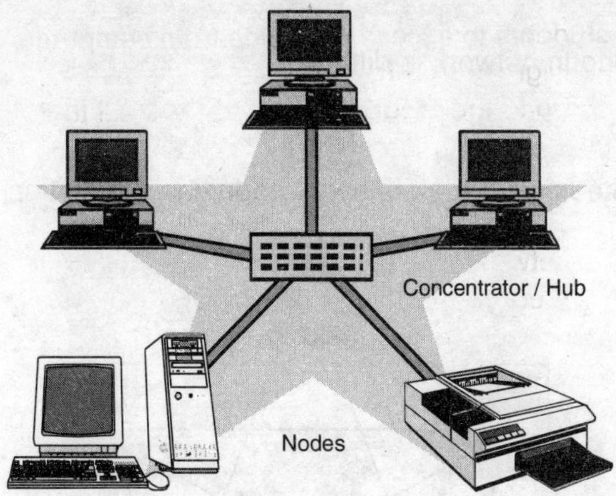

Fig. 1.10.

Disadvantages

(a) The system crucially depend on central node. If the host computer fails, the entire network fails.

2. Ring Topology

Ring consist of ring interfaces conceit to each other by P&P lines. There is no central machine. A node can receive data from one of its two adjacent nodes.

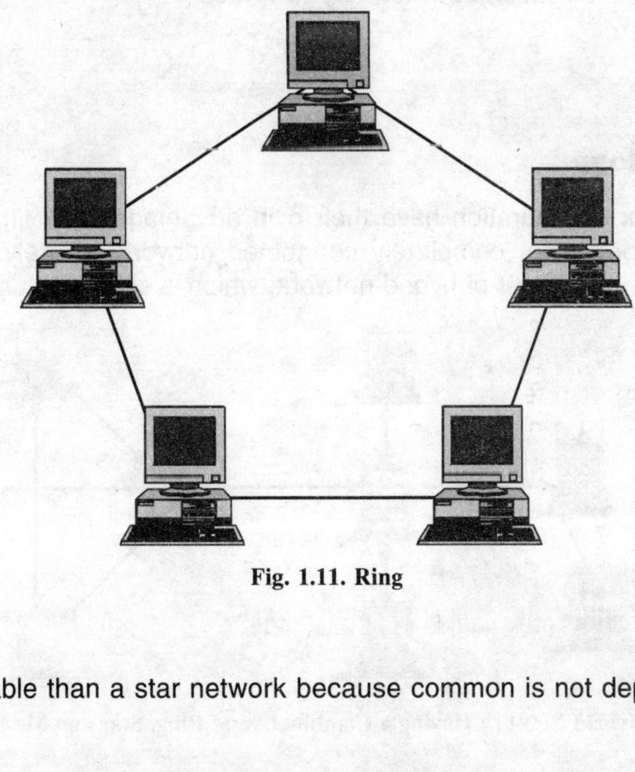

Fig. 1.11. Ring

Advantages

It is more reliable than a star network because common is not dependent on a single host computer.

Disadvantages

Addition of new node in network is difficult.

3. Mesh Topology

It is completely connected network. Here every node has connection with every other node.

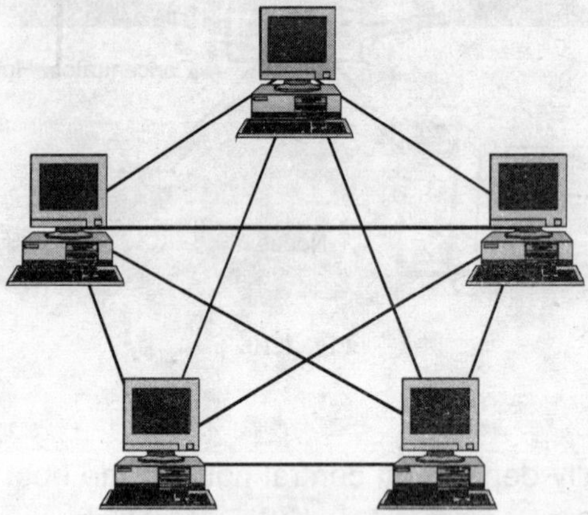

Fig. 1.12. Mesh Topology.

Advantages

(a) Highly reliable.

(b) Very fast communication between any two nodes.

Disadvantages

(a) Very costly.

4. Hybrid Topology

Different network configuration have their own advantages and limitations. Hence in reality a pure star or ring or completely connected network is rarely used instead, an organisation will use same sort of hybrid network, which is simply a combination network.

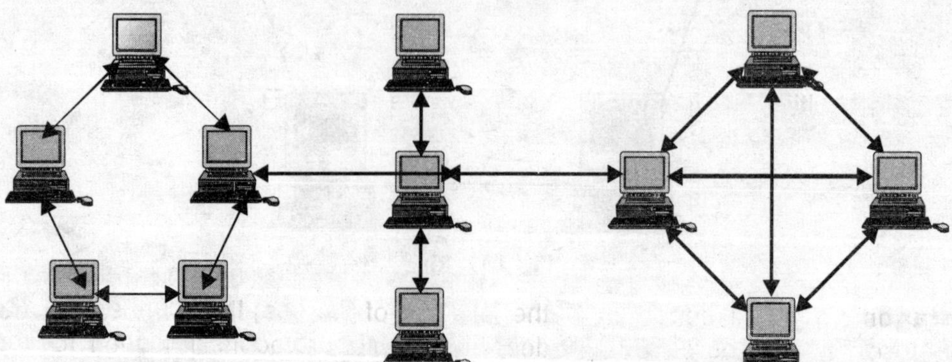

Fig. 1.13. A Hybrid Network Having a Combination of Ring, Star and Mesh Topology.

5. Bus Topology

Physically it is a linear or tree shaped cable

Fig. 1.14. Bus.

- ➢ Slow.
- ➢ Information can be decoded.
- ➢ Main problem is security, password protection is required.

1.9. ARPANET (ADVANCE RESEARCH PROJECTS AGENCY)

History : ARPA was created in response to soviet union's launching sputnik in 1957 and has the mission of advancing technology that might be useful to the military.

Technique : It is a packet switched network consisting of subnet and lost computers.

The subset consist of minicomputer IMP (Interface message processors) connected by transmission lines.

For high reliability, each IMP is connected to atleast two other IMPs the subset was a datagram subnet, so if some lines and IMPs were destroyed, meassages could be automatically routed along alter native paths.

Each node of the network was to consist of IMP and a host, in the same room connected by a short wire A host could send messages of up to 8063 bits to its IMP, which would then break these up to packets of at most 1008 bits and forward them independently towards the destination.

BBN, a consulting firm is combruge modified the ARPANET. They used the IMP having no disks, since moving parts were considered unreliable.

The software was split into two part

1. Subnet
2. Host

The subnet software consisted of the IMP end of the host-IMP connection, the IMP-IMP protocol, and a source IMP to destination IMP protocols designed to improved reliability.

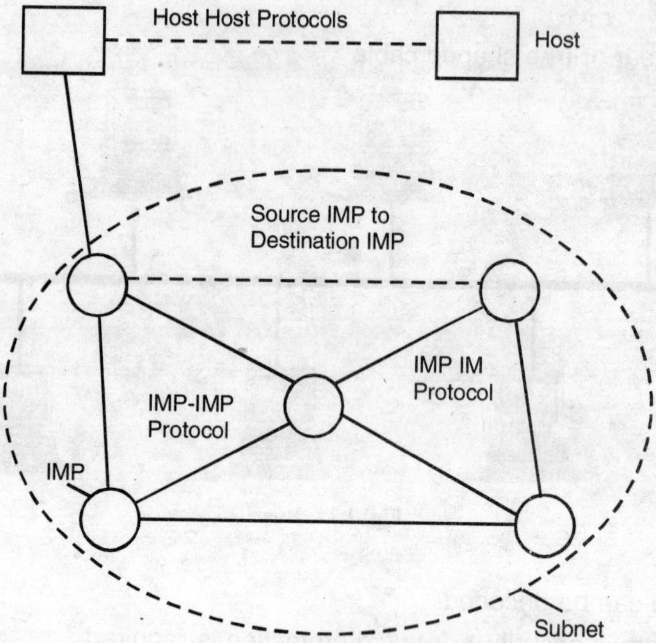

Fig. 1.15. Arpanet Design.

Problem : Outside the subnet, software was needed is host-host protocal and application software later the IMP software was changed to allow terminale to connect directly a special IMP, called TIP (Terminal Interface Processor), which does not pass through a host.

1.10. NSF NET (NATIONAL SCIENCE FOUNDATION NETWORK)

History : To get on the ARPANET, a universities had to have a research contract with the DOD (Department of Defence), which many did not have. This lack of universal access prompted NSF to set up a virtual network, CSNET, Central around a single machine at BBN that supported dial up lines and had connections to the ARPANET and other networks. In 1984 NSF begain designing a high speed successor to the ARPANET that would be open to all university research groups.

Technology : NSF decided to build a backbone network to connect its sin supercomputer centres each supercomputer consist of minicomputer called a *Fuzzball*. The fuzzball were connected with 56 kbps leased lines and formed the subset, the same hardware technology as the ARPANET used.

The software technology was different, the fuzzballs spoko TCP/IP right from the start making it the first TCP/IP WAN.

NSF also funded some regional networks that connected to the backbone to allow users at thousands of uniersities, research labs, libraries to access any of the supercomputer and the communicate with one another the complete network including the backbone and the regional networks was called NSFNET. It connected to the ARPANET through a link between IMP and fuzzball.

1.11. INTERNET [MDU Dec. 2008]

Collection of networks is knows as Internet. A machine is on the internet if the runs the TCP/IP protocol stack, has an internet protocol address and has the ability to send IP packets to all the other machines on the Internet.

Internet had four main application :

1. **E-mail :** The message can be composed, send and received by electronic mail many people get dozens of messages a day and consider it their primary way of interacting with the outside world.
2. **News :** Newsgroups are specialized joruns in which users with a common interest can exchange message.
3. **Remote Login :** Using the Telnet or other programs users anywhere on the internet can log into any other machine on which they have an account
4. **File Transfer :** Using the file transfer protocol program, it is possible to copy files from one machine on the internet to another.

The world wide web (WWW) is one of the internets consisting of client and server computer handling voice, data video pages. Information is stored as electronic pages on the web.

Internet addresses are classified as :

1. Classfull address
2. Classless address (are described in chapter 2).

Uses of Internet

1. **Search Engine :** It can be used to search anything and everything. Most popular search engines are google and yahoo.
2. **Shopping :** You can do online shopping with the help of internet.
3. **Communication/E-mail :** It helps people to communicate either with the use of social networking websites or through e-mails. Chatting is a major use of the internet.
4. **Job Search :** Nowadays, many people search for their jobs online.
5. **Research :** Research papers are present online which helps in the researcher doing a literature review.
6. Stuyding.

1.12. BRIDGES

Briges are the device to connect multiple LAN's. Bridges operates in data link layer.

1.12.1. Why Bridges are Required

1. The bridges are needed where the multiple LAN are used so that interaction between different department can be achieve. For example, universities and corporate department houses.
2. Bridges are being used in the organisation which are geographically spread over a considerable distance in such organisation, it is cheaper to have multiple LAN's connecting together from bridges.

3. Bridges may be used for load sharing.
4. Using the bridges can help to increase the physical distance covered within single LAN network.
5. Reliability of LAN can be achieved by using bridges at strategic points.
6. Bridges can also be used to enhacne the security of an organisation. Here all the frames are passed to bridge and from there they are distributed as per there source and destination address.

1.12.2. Working of Bridge

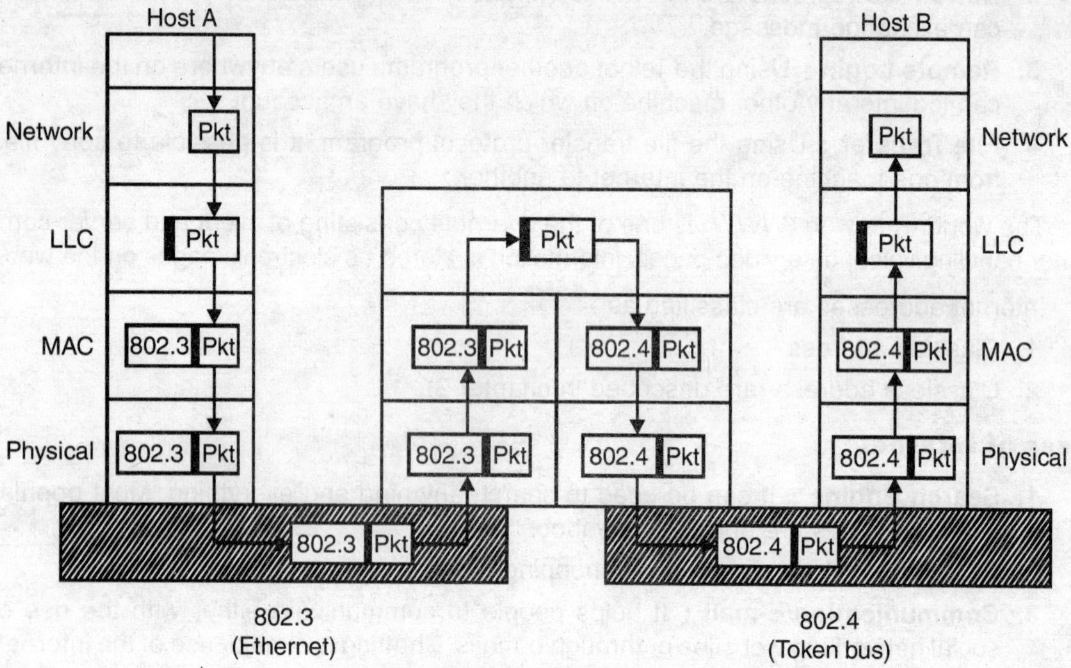

Fig. 1.16. Operation of a LAN Bridge from 802.3 to 802.4.

1.12.3. Working Procedure of Bridge

(a) Host 'A' has a packed to send.

(b) The packet descends into LLC sublayer and acquires an LLC leader.

(c) The it passes into the MAC sublayer and 802.3 leader is prepended to it.

(d) The unit goes out onto the cable and eventually it passed up to the MAC sublayer in bridege where the 802.3 headed is striphed off.

(e) The bare packet with LLC leader is then handed off to LLC sublayer in bridge.

(f) The packet is destined for an 802.4 subnet connected to the bridge, so it works its way down the 802.4 side of the bridge and off it goes.

These bridge are fully transparent. It means you can easily install the bridge without any changes in the hardware and software. These bridge works in promiscuous mode

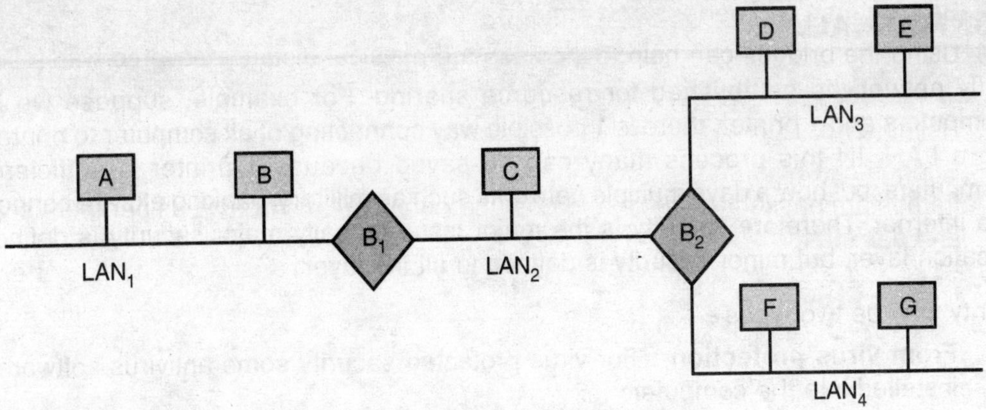

Fig. 1.17. Working Procedure of Bridge.

1.12.4. Types of Bridges

Bridges are of two types :

1. Transparent
2. Source Routing Bridge.

S. No.	Transparent	Source Routing
1.	The bridges are falls of transparent.	These bridges are partially transparent.
2.	They can be easily installed without change in any hardware.	Here we required the newly hardware and software to install the 6 bridge.
3.	They work in promiscuous mode means if the frame is to be transmitted within same LAN, then also it will be acquired by bridge. The bridge then will decide to discard or transmit it on other LAN.	These bridges do not work in promiscuous mode.
4.	We required backward learning algorithm.	We required here discovery frame algorithm.
5.	Complexity is defined in the bridge.	Complexity is defined in the host.
6.	For failure of packets bridge is responsible.	Host is responsible.

1.12.5. Problems With Bridges

1. If interconnected LANs do not run at same data rate.
2. When a long frame must be forwarded onto a LAN that cannot accept it.
3. Bolteneck problem, when timer goes off.

1.12.6. Advantages of Bridges

1. Security
2. Portability.

1.13. FIREWALL

Initially network is established for resource sharing. For example, suppose we have 10 computers and 1 printer, there is 1 possible way connecting of all computer to printers to make a LAN. IN this process many can be saved because 1 printer is sufficient for 10 computers, but now a days multiple networks such as millitary, banking etc. are connected to the internet. Therefore security is the major issue actually major security is defined in application layer, but minor security is defined in all the layer.

Security may be two types :

1. **From virus protection :** For virus protected security some antivirus software are installed into the computer.
2. **From unauthorized user :** Unauthorized user cannot access to the computer using the cryptology, but it is the issue of application layer.

In the internetworking security is provided by the Firewall. Firewall is the collection of special software (E.g., antivirus software) and hardware.

1.13.1. Construction of Firewall

Two filtering power are connected by application gatewall.

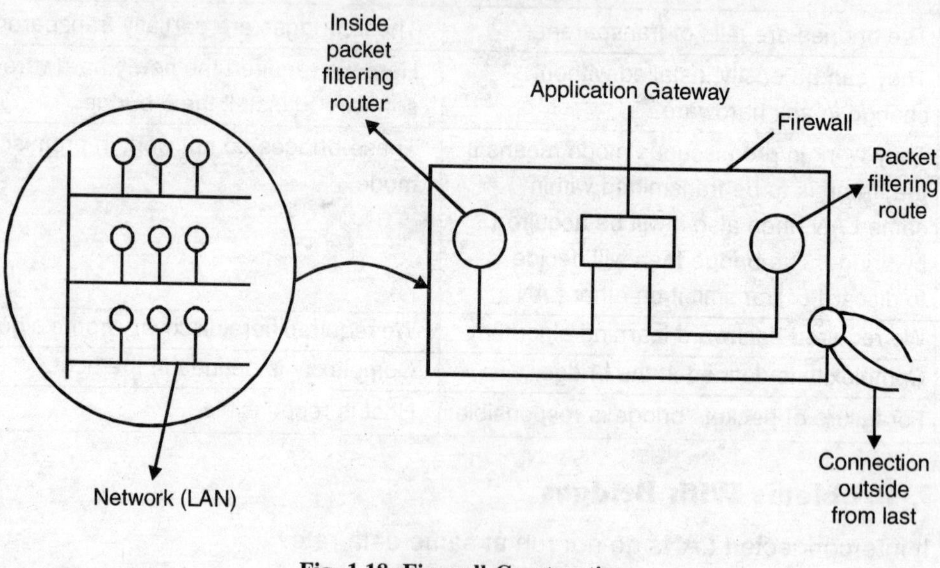

Fig. 1.18. Firewall Construction.

It contains two filtering routers outsider filtering router and inside filtering router and application gateway.

1. Outside Filtering Router : It is responsibility for security if packet is coming from outside the LAN (means unauthorized user). Actually filtering router are special router, which contains filtering software.

Filtering means security of secret message from unauthorized user.

2. Inside Filtering Router : It is responsible for security of packets going from LAN to outside. It also contains filtering software.

3. Application Gateway.

1.14. INTERNETWORKING

Network layer has three design issues :

1. Routing decision
2. Congestion control
3. It allows heterogeneous networks to be inter connected (Inter networking).

Internetworking in the heterogeneous network we have the different types of system, each system contains different hardware and software configuration but in the communication media different type of configuration is not allowed. It is impossible every machine contains same type of configuration. We can connect different type of system (hetergeneous system) through the internet working.

DO YOU KNOW

Short form of internet work is internet.

TIPS

Internet working is the collections of different type of network.

There may be following type of internet working

(a) LAN – LAN
(b) LAN – WAN
(c) LAN – WAN – LAN
(d) WAN – WAN

Following hardware and software are required for internet working.

1. Repeaters : Actually internet working is spread in large geographical area. Therefore signal cannot reach to the distant network. It can be reach to the distant network. It can be reach to the distant network using repeaters. Repeaters are installed at energy few km so that it can maintain the original signal.

2. Multiprotocol Routers : Routers are used for taking a decision for packet forwarding to the destination. It is used in the network layer.

3. Transport Gateway : Transport gateway are used for the connecting of network to the transport layer. (used in transport layer).

4. Application Gateway : Application gateway are used in application layer. In the application layer it is used for formatting of the message and message is appear same in all the network.

1.14.1. Difference between Routers, Gateways and Bridges

S. No.	Bridges	Router	Gateways
1.	Used to connect multiple LAN's.	Used to connect network that may or may not be similar.	Used to connect 2 distinguish LAN's/WAN's/Man's
2.	Act as an address filter is relay of frames between same kind of networks	Provides connectivity between 2 LANs or 2WANs over large geographical distance.	Used toconnect any type of the networks with any type of network through software.
3.	They operate in DLL.	They operate at N/W layer.	It operate in application layer gateway.
4.	Bridge is simple hardware device able to execute specific task.	A router routes protocols between potentially different N/Ws.	Gateway can accept and transfer data packets between different protocol network is they are protocol conventors.
5.	They cannot take decision for routing of packets.	More sophisticated and uses routing algo to find best available path for data transmission.	Are generally so ware installed within a router.

1.15. NETWORK PARTIONING HARDWARE

Network partitioning involves dividing a single backbone into multiple segments, joined by some piece of hardware that forwards packets. There are multiple types of these devices:

- Repeaters
- Hubs
- Bridges
- Switches
- Routers, and
- Gateways.

These terms are sometimes used interchangeably although each device has a specific set of policies regarding packet forwarding, protocol filtering, and transparency on the network:

Repeaters

A repeater joins two segments at the physical layer. It is a purely electrical connection, providing signal amplification and pulse "clean up" functions without regard for the semantics of the signals. Repeaters are primarily used to exceed the single-cable length limitation in networks based on bus topologies, such as 10Base5 and 10Base2. There is a maximum to the number of repeaters that can exist between any two nodes on the same network, keeping the minimum end-to-end transit time for a packet well within the Ethernet specified maximum time-to-live. Because repeaters do not look at the contents of packets (or packet fragments), they pass collisions on one segment through to the other, making them of little use to relieve network congestion.

TIPS

Repeaters are also called regenerator. It is an electronoc device that operates onlyh at physical layer.

Hubs

A hub joins multiple hosts by acting as a wiring concentrator in networks based on star topologies, such as 10BaseT. A hub has the same function as a repeater, although in a different kind of network topology. Each computer is connected, typically over copper, to the hub, which is usually located in a wiring closet. The hub is purely a repeater: it regenerates the signal from one set of wires to the others, but does not process or manage the signal in any way. All traffic is forwarded to all machines connected to the hub.

Bridges

Bridges function at the data link layer, and perform selective forwarding of packets based on their destination MAC addresses. Some delay is introduced into the network by the bridge, as it must receive entire packets and decipher their MAC-layer headers. Broadcast packets are always passed through, although some bridge hardware can be configured to forward only ARP broadcasts and to suppress IP broadcasts such as those emanating from *ypbind.*

Intelligent or learning bridges glean the MAC addresses of machines through observation of traffic on each interface. "Dumb" bridges must be loaded with the Ethernet addresses of machines on each network and impose an administrative burden each time the network topology is modified. With either type of bridge, each new segment is likely to be less heavily loaded than the original network, provided that the most popular inter-host virtual circuits do not run through the bridge.

TIPS

These operate both in physical and data link layers of LAN of same type.

Switches

You can think of a switch as an intelligent hub having the functionality of a bridge. The switch also functions at the data link layer, and performs selective forwarding of packets based on their destination MAC address. The switch forwards packets only to the intended port of the intended recipient. The switch "learns" the location of the various MAC addresses by observing the traffic on each port. When a switch port receives data packets, it forwards those packets only to the appropriate port for the intended recipient. A hub would instead forward the packet to all other ports on the hub, leaving it to the host connected to the port to determine its interest in the packet. Because the switch only forwards the packet to its destination, it helps reduce competition for bandwidth between the hosts connected to each port.

Routers

Repeaters, hubs, bridges, and switches divide the network into multiple distinct physical pieces, but the collection of backbones is still a single *logical* network. That is, the IP network number of all hosts on all segments will be the same. It is often necessary to divide a network logically into multiple IP networks, either due to physical constraints (i.e., two offices that are separated by several miles) or because a single IP network has run out of host numbers for new machines.

Multiple IP networks are joined by routers that forward packets based on their source and destination IP addresses rather than 48-bit Ethernet addresses. One interface of the router is considered "inside" the network, and the router forwards packets to the "outside" interface. A router usually corrals broadcast traffic to the inside network, although some can be configured to forward broadcast packets to the "outside" network. The networks joined by a router need not be of the same type or physical media, and routers are commonly used to join local area networks to point-to-point long-haul internetwork connections. Routers can also help ensure that packets travel the most efficient paths to their destination. If a link between two routers fails, the sending router can determine an alternate route to keep traffic moving. You can install a dedicated router, or install multiple network interfaces in a host and allow it to route packets in addition to its other duties contains a detailed description of how IP packets are forwarded and how routes are defined to Unix systems.

Gateways

At the top-most level in the network protocol stack, a gateway performs forwarding functions at the application level, and frequently must perform protocol conversion to forward the traffic. A gateway need not be on more than one network; however, gateways are most commonly used to join multiple networks with different sets of native protocols, and to enforce tighter control over access to and from each of the networks.

Replacing an Ethernet hub with a Fast Ethernet hub is like increasing the speed limit of a highway. Replacing a hub with a switch is similar to adding new lanes to the highway. Replacing an Ethernet hub with a Fast Ethernet switch is the equivalent of both improvements, although with a higher cost.

TIPS

They relay packets among network that have different protocols. They accept a packet formatted for one protocol and convert it to a packet formatted for another protocol before forwarding it.

HIGHLIGHTS

1. Interconnection of autonomous computer is called computer networking.
2. Interconnection means information can be exchanged.
3. Advantages of computer networking :
   - ➢ File sharing
   - ➢ Resource sharing

- ➢ Increased storage capacity
- ➢ Increased cost efficiency
- ➢ Flexible access
- ➢ Centralise software management.

4. Disadvantages of computer networking :
- ➢ Security issue
- ➢ Virus threat
- ➢ Network halt
- ➢ Expensive setup
- ➢ Dependency on main file server.

5. Computer networking can be of two types :
- ➢ Broadcast
- ➢ P2P (Point to Point)

6. In broadcast network there is single communication channel on which all nodes operates.

7. In P2P Network, there are multiple communication channel.

8. Computer network on the basis of scale is of three types :
- ➢ LAN
- ➢ WAN
- ➢ MAN

9. Network topology means the shape or architecture that is used to join the network.

10. Various network topology are possible :
- ➢ Star topology
- ➢ Ring topology
- ➢ Mesh topology
- ➢ Bus topology
- ➢ Hybrid topology

11. The internet is a global system of interconnected computer networks that use TCP/IP suite.

12. Bridges are device to connect multiple LAN.

13. Bridges are of two types :
- ➢ Transparent.
- ➢ Source routing.

14. In the internetworking security is provided by the firewall.

15. Firewall consists of
- ➢ Outside filtering router
- ➢ Inside filtering router
- ➢ Application gateway.

16. Internetworking allows heterogeneous network to be interconnected.

17. Internet working is of following types :
- ➢ LAN – LAN
- ➢ LAN – WAN
- ➢ LAN – WAN – LAN
- ➢ WAN – WAN.

18. Various network partitioning hardware are :
- ➢ Repeaters
- ➢ Hubs
- ➢ Bridges
- ➢ Switch
- ➢ Router

QUESTIONS WITH ANSWERS

Q. 1. What are differences between Virtual Circuit and Datagram.

Ans.

S.No.	Virtual Circuit	Datagram
1.	They use connection from source to destination.	Connection set up is not required.
2.	While using virtual circuit, each packet contains a short virtual circuit number. There is no need of fall source-destination address.	While working with datagram, each packet must contain the full source and destination address, as each packet is routed independently of its previous packet.
3.	Congestion control is not difficult in virtual circutting if enough buffers can be allocated in advance.	Congestion control is difficult in datagrams
4.	A preplanned route is established before any packet are sent, all packets follow this route is route established for entire conversation.	Each packet is routed independently is route established for each packet.
5.	If router fails, all vireval circuits that passed through the failed router are terminated.	If a router fails, then packet can be routed from other route, so there is no less, except for packets lost during the crash. So it is more reliable.
6.	Virtual circuit uses less memory than datagrams as each packet contain a short virtual circuit number instead of full address.	Datagrams uses more bandwidth than virtual circuit as they have to keep full source and destination address.
7.	For long messages, the virtual circuit technique is superior.	For short messages, datagram will be faster than virtual switching. It is not goal for long message as processing for each datagram at each node may be longer than for Vc.

Q. 2. What are the different types of networking and internetworking devices.

Ans. **Repeater :** Also called a regenerator, it is an electronic device that operates only at physical layer. It receives the signal in the network before it becomes weak, regenerates the original bit pattern and puts the refreshed copy back in to the link.

Bridges : These operate both in the physical and data link layers of LANs of same type. They divide a larger network in to smaller segments. They contain logic that allow them to keep the traffic for each segment separate and thus are repeaters that relay a frame only the side of the segment containing the intended recipent and control congestion.

Routers : They relay packets among multiple interconnected networks (i.e. LANs of different type). They operate in the physical, data link and network layers. They contain software that enable them to determine which of the several possible paths is the best for a particular transmission.

Gateways : They relay packets among networks that have different protocols (e.g. between a LAN and a WAN). They accept a packet formatted for one protocol and convert it to a packet formatted for another protocol before forwarding it. They operate in all seven layers of the OSI model.

Q. 3. What is difference between broadband and baseband transmission ?

Ans. In base band transmission we transmit digital signal without converting it into analog.here a low pass channel is used.

In broad band transmission we transmit digital signal by converting it into analog.Here a band pass channel is used

Q. 4. What is a VLAN? What does VLAN provide?

Ans. A technology called VLAN (Virtual LAN broadcast domains logically segmented on an Ethernet switch) trunking that was once primarily the domain of network switches has now trickled down to the rest of the Data Center to address these issues. Now it is possible for these multi-homing devices to be multi-homing in function without the need for multiple physical network adapters and the additional infrastructure associated with them. VLAN trunking allows a single network adapter to behave as ?n? number of virtual network adapters, where ?n? has a theoretical upper limit of 4096 but is typically limited to 1000 VLAN network segments. In the case where a single gigabit Ethernet adapter is trunked in place of using multiple FastEthernet adapters, higher performance at a lower cost while increasing flexibility can be achieved.

Q. 5. What is a difference between switch and Hub?

Ans. **Hub :** (i) Hub is a layer-1 device,in this data transmission in the form of bits.

(ii) It is a multiple-port repeater.any signals send via the hub is transmitted to all the the ports on the hub.

Switch : (i) Switch is a layer-2 device,in this data transmission in the form of frames.

(ii) It transmit data only to the destination port.

Q. 6. What are the types of Transmission media?

Ans. Signals are usually transmitted over some transmission media that are broadly classified in to two categories:

Guided Media : These are those that provide a conduit from one device to another that include twisted-pair, coaxial cable and fiber-optic cable. A signal traveling along any of these media is directed and is contained by the physical limits of the medium. Twisted-pair and coaxial cable use metallic that accept and transport signals in the form of electrical current. Optical fiber is a glass or plastic cable that accepts and transports signals in the form of light.

Unguided Media : This is the wireless media that transport electromagnetic waves without using a physical conductor. Signals are broadcast either through air. This is done through radio communication, satellite communication and cellular telephony

REVIEW QUESTIONS

1. What is computer network. What are the advantages of computer networking.
2. What are the disadvantages of computer networking ?
3. What is the difference between broadcast and P2P network ?
4. Discuss various network partitioning hardware.
5. What are different types of bridges.
6. What is internet ?
7. What is inter working ?
8. Differentiate LAN, WAN, MAN.
9. Discsu ARPANET.
10. Discuss NSFNET.
11. What is the difference between broadcasting and multicasting.

❑❑❑

TCP/IP Protocol Architecture

Chapter 2

INSIDE THIS CHAPTER

2.1. OSI MODEL(OPEN SYSTEM INTERCONNECTION)

It was established in 1947, by the International Standards Organization (ISO).ISO is a multinational body dedicated to worldwide agreement on international standards. An ISO standard that covers all aspects of network communications is the Open Systems Interconnection (OSI) model. It was first introduced in the late 1970s.

OSI model consists of seven layers:

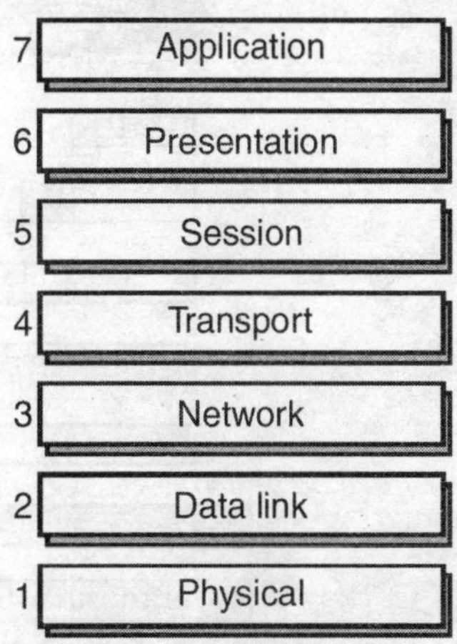

Fig. 2.1.

DO YOU KNOW

ISO is the organization and OSI is a model.

2.1.1. OSI layers

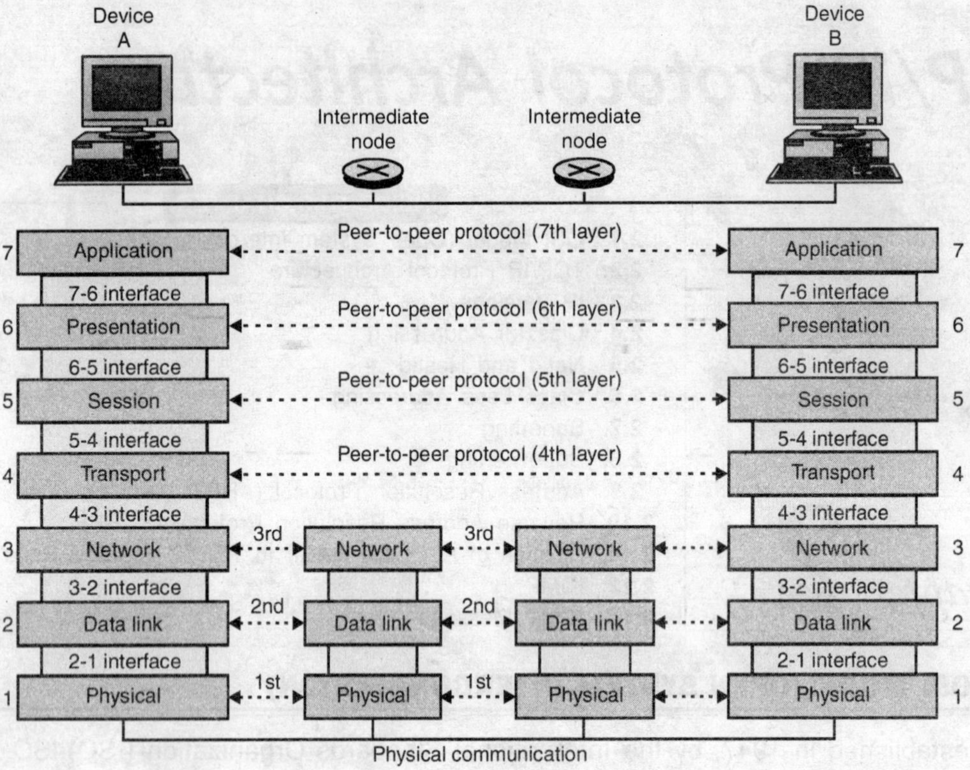

Fig. 2.2.

An Exchange Using the OSI Model

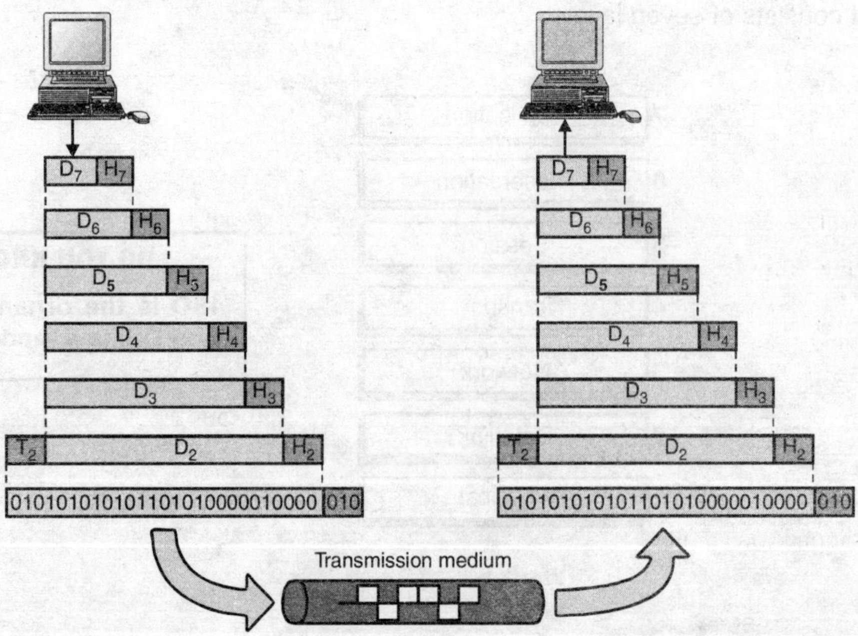

Fig. 2.3.

2.1.2. Functions of Each Layer

1. Physical Layer

The physical layer is responsible for the movement of individual bits from one hop (node) to the next.

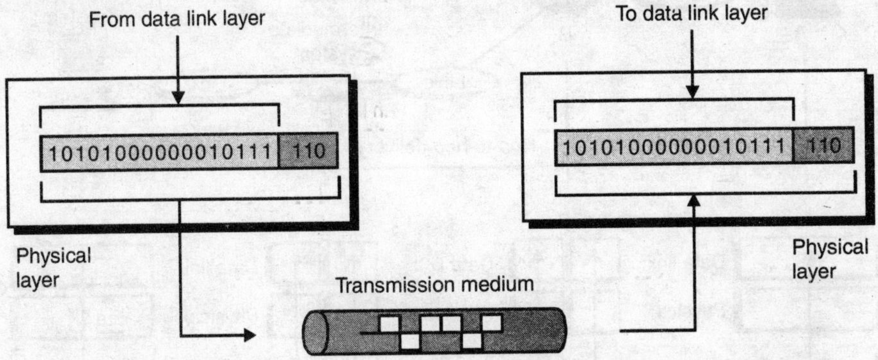

Fig. 2.4.

2. Data Link Layer (DLL)

The data link layer is responsible for moving frames from one hop (node) to the next.

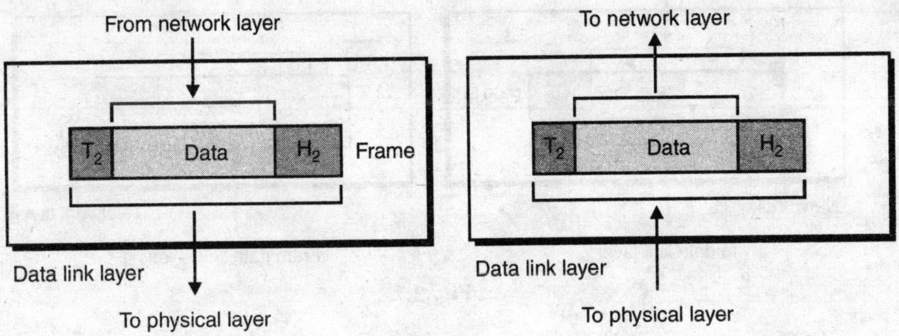

Fig. 2.5.

- It is responsible for framing.
- Flow control
- Error control.

3. Network Layer

The network layer is responsible for the delivery of individual packets from the source host to the destination host. This layer, also known as the Ethernet layer, is responsible for machine-to-machine communications. It determines the path a transmission must take, based on the receiving machine's IP address. Besides message routing, it also translates from logical to physical addresses and provides error detection. In short it provides

- Routing
- Congestion control
- Internet working
- IP addresses

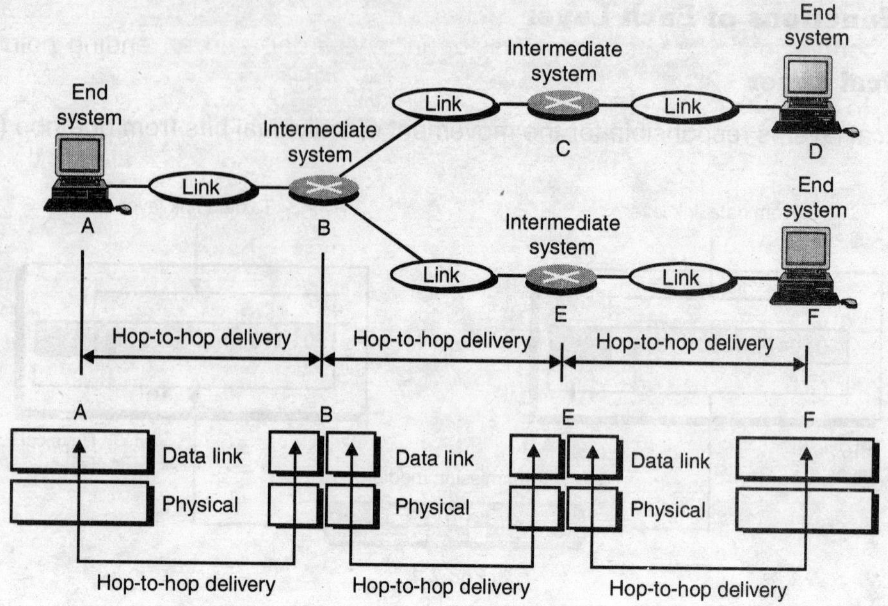

Fig. 2.6.

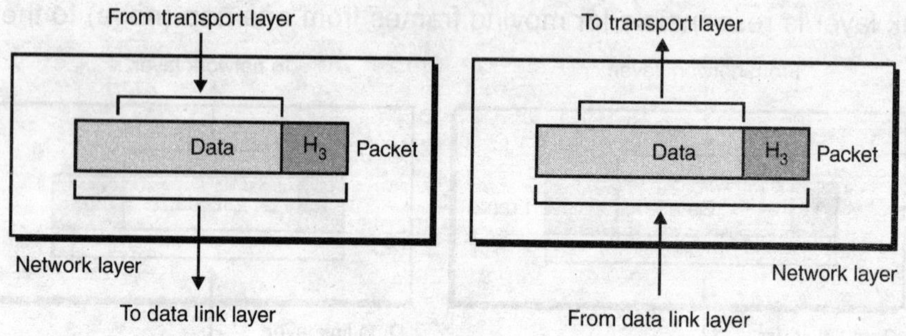

Fig. 2.7.

4. Transport Layer

The transport layer is responsible for the delivery of a message from one process to another. The transport layer controls the flow of data on the network and assures that received and transmitted data are identical. TCP/IP or UDP may be used to enable communications between application programs running on separate machines. In short it provides

- Flow control
- Error control
- Congestion control
- Port address
- Connection management.

It uses two two protocol.

TCP Protocol (Transmission Control Protocol)

TCP enables applications to communicate with each other as though connected by a physical circuit. TCP sends data in a form that appears to be transmitted in a character-by-character fashion, rather than in discrete packets. This transmission consists of a starting point, which opens the connection, the entire transmission in byte order, and an ending point, which closes the connection.

TCP makes an attempt to confirm that a packet has reached its destination by establishing an end-to-end connection between sending and receiving hosts. TCP is therefore considered a "reliable, connection-oriented" protocol.

TIPS

TCP is reliable connection oriented protocol.

UDP Protocol (User Datagram Protocol)

UDP provides datagram delivery service. It does not provide any means of verifying that connection was ever achieved between receiving and sending hosts. Because UDP eliminates the processes of establishing and verifying connections, applications that send small amounts of data often use it rather than TCP. UDP is a "connectionless" protocol.

Source-to-Destination Delivery

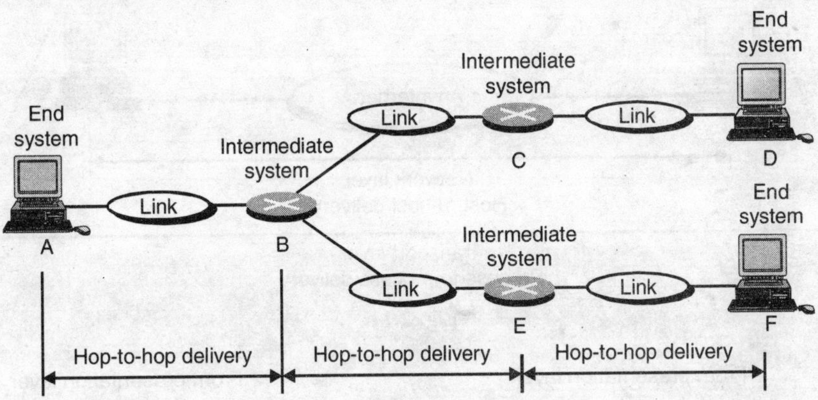

Fig. 2.8.

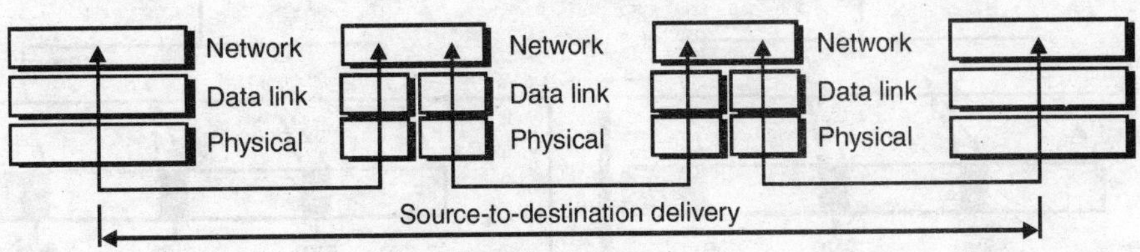

Fig. 2.9.

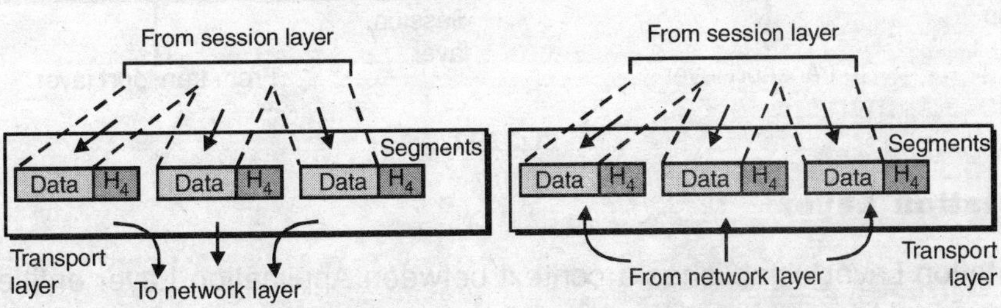

Fig. 2.10.

5. Session Layer

It is responsible to manage the sessions.

- ➢ The Session Layer controls the dialogues/connections (sessions) between computers.
- ➢ It establishes, manages and terminates the connections between the local and remote application.
- ➢ It provides for full-duplex, half-duplex, or simplex operation, and establishes checkpointing, adjournment, termination, and restart procedures.

The OSI model made this layer responsible for "close" of sessions, which is a property of TCP, and also for session checkpointing and recovery, which is not usually used in the Internet Protocol Suite. The Session Layer is commonly implemented explicitly in application environments that use remote procedure calls (RPCs).

It provides reliable process-to-process delivery of a message

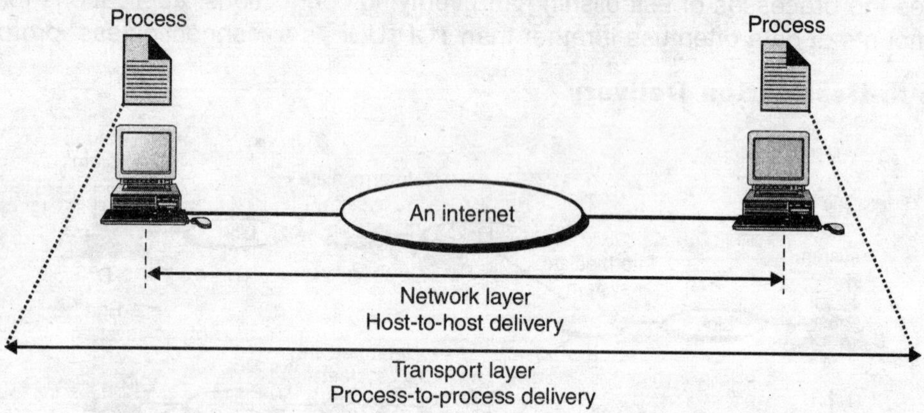

Fig. 2.11.

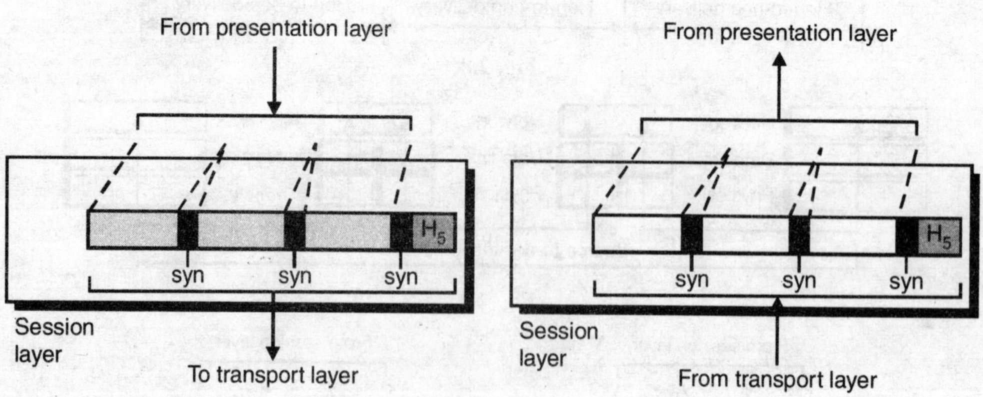

Fig. 2.12.

6. Presentation Layer

The Presentation Layer establishes a context between Application Layer entities, in which the higher-layer entities can use different syntax and semantics, as long as the Presentation

Service understands both and the mapping between them. The presentation service data units are then encapsulated into Session Protocol Data Units, and moved down the stack.

This layer provides :

➢ Independence from differences in data representation (e.g., encryption) by translating from application to network format, and vice versa.

➢ The presentation layer works to transform data into the form that the application layer can accept.

➢ This layer formats and encrypts data to be sent across a network, providing freedom from compatibility problems. It is sometimes called the syntax layer.

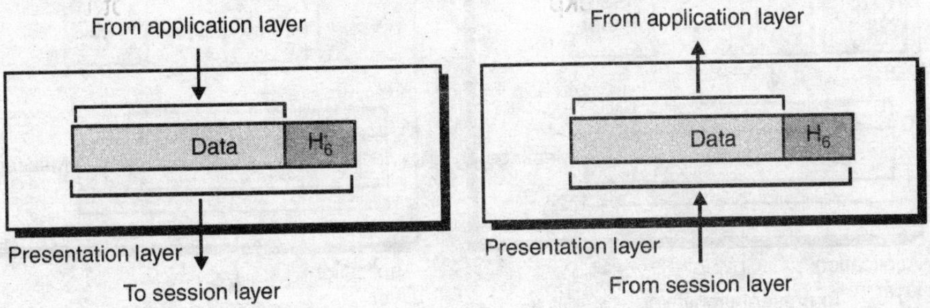

Fig. 2.13.

7. Application Layer

The application layer defines standard Internet services and network applications that anyone can use. These services work with the transport layer to send and receive data. There are many applications layer protocols, some of which you probably already use. Some of the protocols include:

➢ Standard TCP/IP services such as the FTP, TFTP, and TELNET commands

➢ UNIX "r" commands, such as rlogin and rsh

➢ Name services, such as NIS+ and Domain Name System (DNS)

➢ File services, such as the NFS system

TIPS

The application layer is the OSI layer closest to the end user, which means that both the OSI application layer and the user interact directly with the software application.

This layer interacts with software applications that implement a communicating component. Such application programs fall outside the scope of the OSI model. Application layer functions typically include identifying communication partners, determining resource availability, and synchronizing communication. When identifying communication partners, the application layer determines the identity and availability of communication partners for an application with data to transmit. When determining resource availability, the application layer must decide whether sufficient network resources for the requested communication exist. In synchronizing communication, all communication between applications requires cooperation that is managed by the application layer.

> **TIPS**
>
> **Some examples of application layer implementations include Telnet, File Transfer Protocol (FTP), and Simple Mail Transfer Protocol (SMTP).**

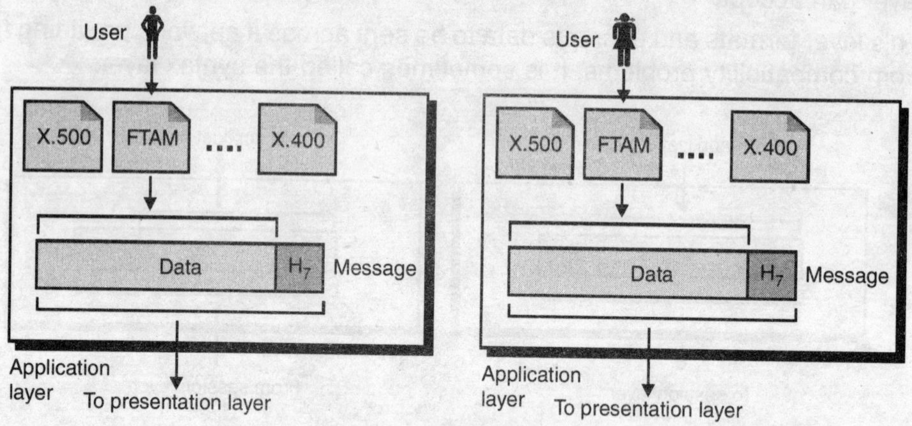

Fig. 2.14.

Key Points About Layer

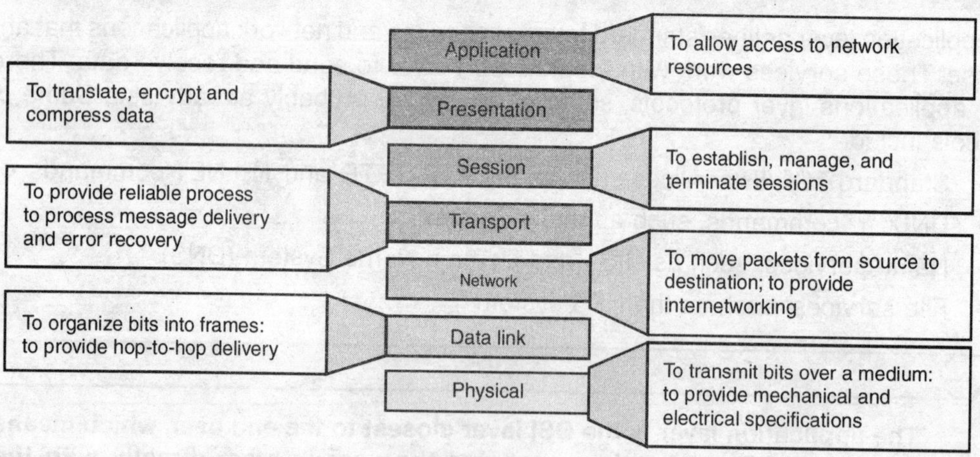

Fig. 2.15.

2.2. TCP/IP PROTOCOL ARCHITECTURE [MDU May 2008; Dec. 2006; May 2006]

The TCP/IP protocol suite is made of five layers :

1. Physical Layer
2. Data link layer
3. Network layer
4. Transport layer
5. Application.

The first four layers provide physical standards, network interface, internetworking, and transport functions that correspond to the first four layers of the OSI model. The three topmost layers in the OSI model, however, are represented in TCP/IP by a single layer called the application layer.

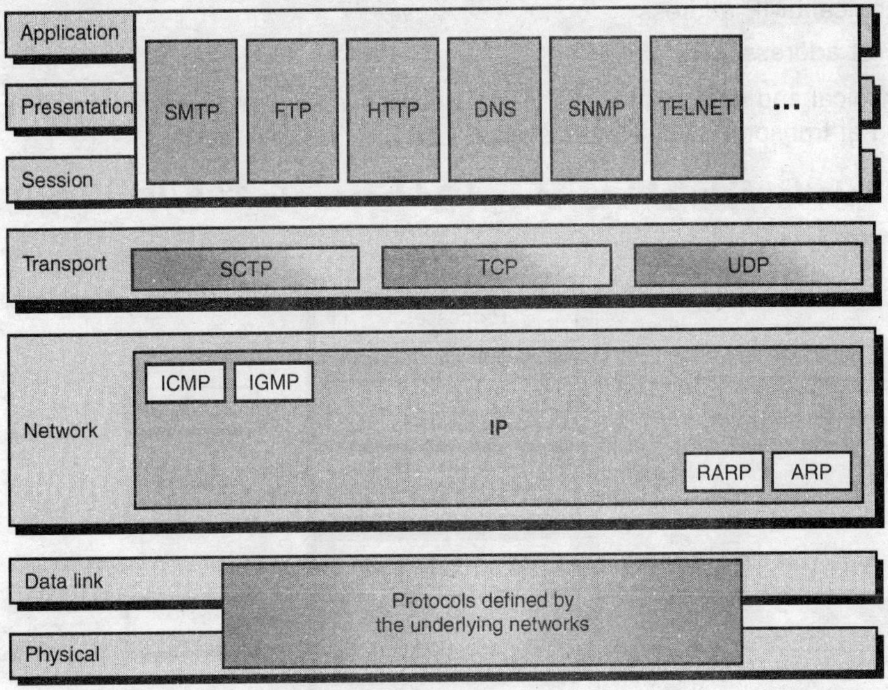

Fig. 2.16.

Difference between OSI Model and TCP IP Model

S. No.	OSI	TCP/IP
1.	The OSI model was devired before the protocols were invented.	In TCP/IP modes the protocols came first and model was based on the protocols.
2.	The OSI model has seven layers.	TPC-IP model has four layer.
3.	The OSI model supports both conection less and connection oriented communication in the network layer, but only connection -oriented communication in the transport layer.	In TCP/IP model has, network layer is connection less and transport layer can be connection less or connection oriented. Communication in the transport layer.
4.	OSI model distinguishes the concept of service, *interface* and *protocols* carefully. The protocols in OSI model cas be rep-laced relationely easily as the technology changes.	This model does not cleary distinguish the concepts of service, interface and protocol. As TCP/IP model does not clearly distin-guishes the protocols, service, interface, so it is not suitable for designing networks using new technologies.
5.	OSI model clearly distinguish the physical layer and data link layer.	The TCP/IP model does not distinguish the physical and data link layers. There are totally different.

2.2.1. Addresses in TCP/IP

Three different levels of addresses are used in an internet using the TCP/IP protocols:

- Physical or link address,
- Logical or IP address
- Port address

At physical and data link layer physical address is used at network layer IP address is used, and at transport layer port address is used

2.2.2. Relationship of Layers and Addresses in TCP/IP

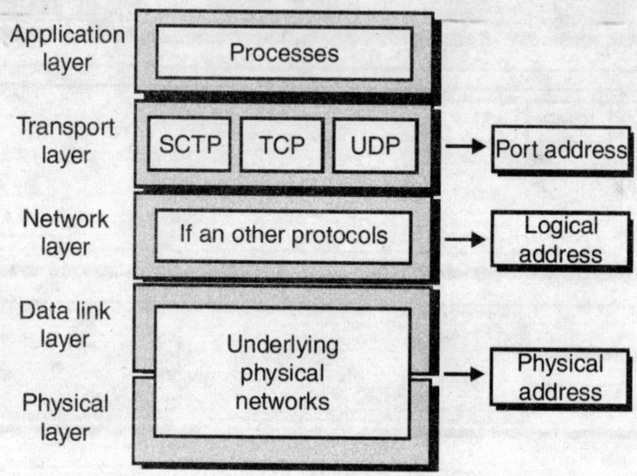

Fig. 2.17. Relationship of Layers.

TIPS

Network layer is responsible for IP address. Physical layer is responsible for physical address. Transport layer is responsible for port address.

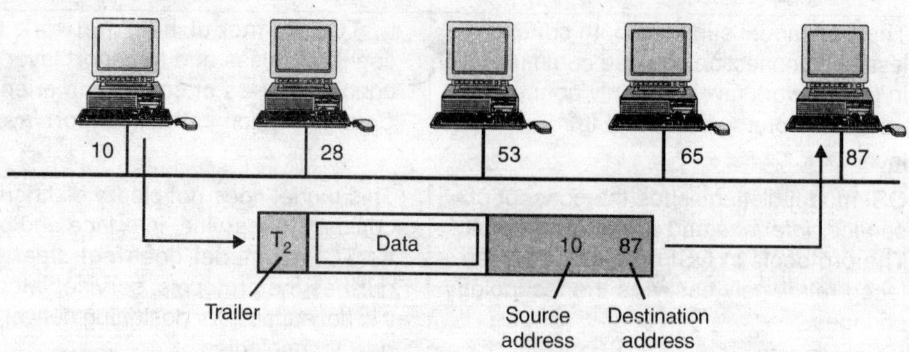

Fig. 2.18. Physical Layer.

Example 1. In following Fig. 2.18, a node with physical address 10 sends a frame to a node with physical address 87. The two nodes are connected by a link. At the data link level this frame contains physical (link) addresses in the header. These are the only addresses needed. The rest of the header contains other information needed at this level. The trailer usually contains extra bits needed for error detection.

Physical Layer

Example 2. Most local area networks use a 48-bit (6 bytes) physical address written as 12 hexadecimal digits, with every 2 bytes separated by a colon as shown below:

07:01:02:01:2C:4B

A 6-byte (12 hexadecimal digits) physical address.

> **TIPS**
>
> **Logical address is also called IP address.**

Example 3. In following figure, we want to send data from a node with network address A and physical address 10, located on one LAN, to a node with a network address P and physical address 95, located on another LAN. Because the two devices are located on different networks, we cannot use link addresses only; the link addresses have only local jurisdiction. What we need here are universal addresses that can pass through the LAN boundaries. The network (logical) addresses have this characteristic.

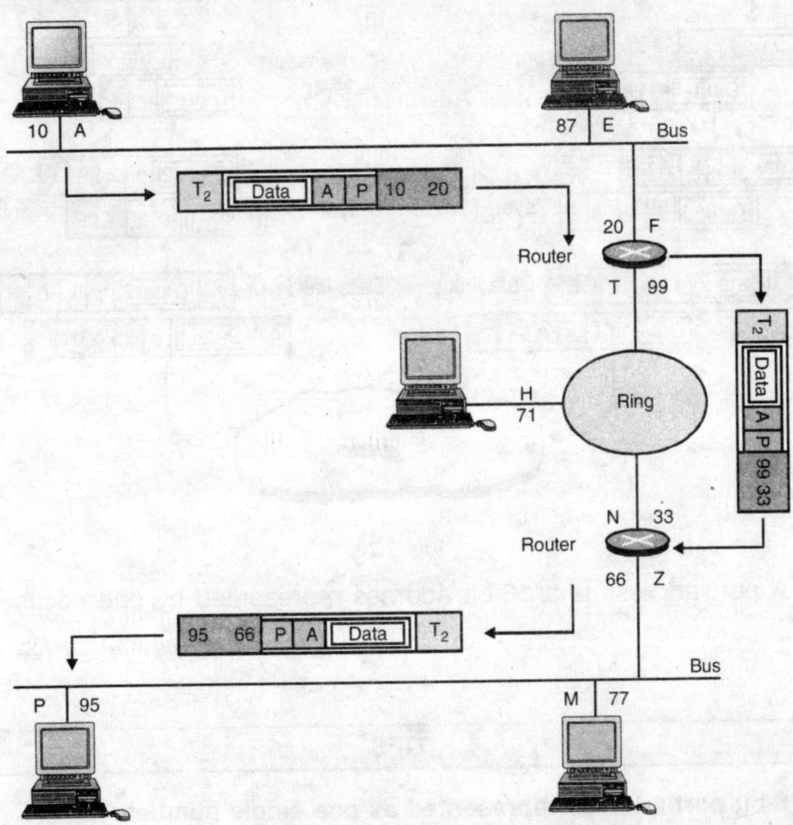

Fig. 2.19.

Example 4. An Internet address (in IPv4) is 32 bits in length, normally written as four decimal numbers, with each number representing 1 byte. The numbers are separated by a dot. Below is an example of such an address.

```
132.24.75.9
```

An internet address in IPv4 in decimal numbers.

Example 5. Following Figure 2.20 shows an example of transport layer communication. Data coming from the upperlayers have port addresses j and k (j is the address of the sending process, and k is the address of the receiving process). Since the data size is larger than the network layer can handle, the data are split into two packets, each packet retaining the service-point addresses (j and k). Then in the network layer, network addresses (A and P) are added to each packet. The packets can travel on different paths and arrive at the destination either in order or out of order. The two packets are delivered to the destination transport layer, which is responsible for removing the network layer headers and combining the two pieces of data for delivery to the upper layers.

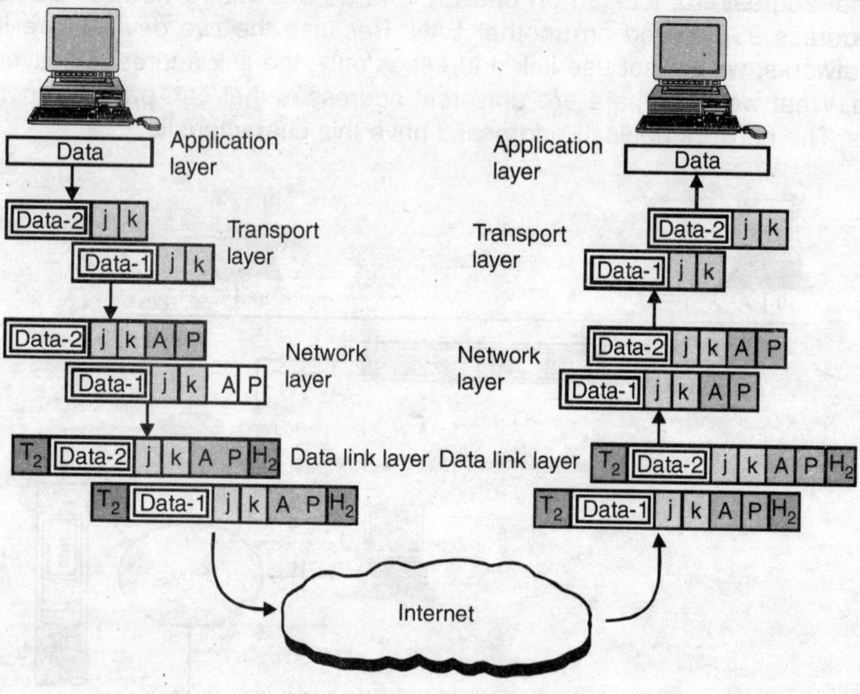

Fig. 2.20.

Example 6. A port address is a 16-bit address represented by one decimal number as shown below.

```
753
```

TIPS

A 16-bit port address represented as one single number.

2.3. IP VERSIONS

IP became the official protocol for the Internet in 1983. As the Internet has evolved, so has IP. There have been six versions since its inception. Three versions are main version

1. Version 4
2. Version 5
3. Version 6

IP Addresses

[MDU Dec. 2008; May 2008]

The identifier used in the IP layer of the TCP/IP protocol suite to identify each device connected to the Internet is called the Internet address or IP address.

OR

An IP address is a 32-bit address that uniquely and universally defines the connection of a host or a router to the Internet.

There are three common ways in which IP address can be represented.

(1) There is the binary notation which uses the base two number system to represent numbers.

(2) There is the decimal notation which uses the base ten number system to represent numbers.

(3) There is the hexadecimal notation which uses the base sixteen number system to represent numbers.

DO YOU KNOW

An IP address is a 32-bit (4-bytes) address.

Example 1. MDU university IP address is 142.190.23.180. This IP address consists of four bytes. The first byte has the value 142. The second byte has the value of 190. The third byte has the value 23, and the fourth byte has the value 18

IP addresses are unique in the sense that each address defines one, and only one, connection to the Internet. Two devices on the Internet can never have the same address.

Notations of IP Addresses

DO YOU KNOW

IP addresses are unique.

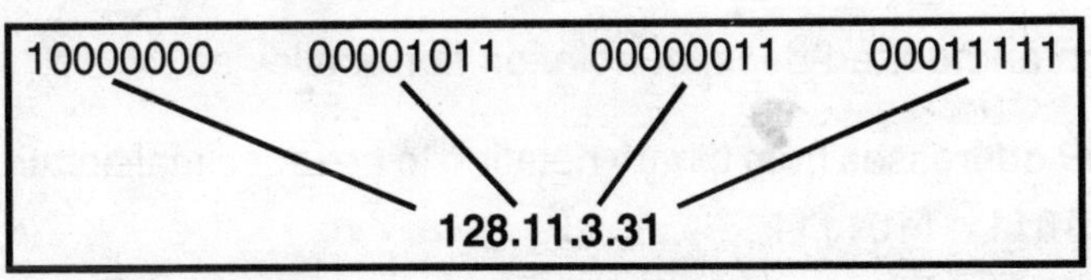

Fig. 2.21.

Example 1. Change the following IP addresses from binary notation to dotted-decimal notation.

(a) 10000001 00001011 00001011 11101111

(b) 11000001 10000011 00011011 11111111

(c) 11100111 11011011 10001011 01101111

(d) 11111001 10011011 11111011 00001111

Solution. We replace each group of 8 bits with its equivalent decimal number and add dots for separation :

(a) 129.11.11.239
(b) 193.131.27.255
(c) 231.219.139.111
(d) 249.155.251.15

> **DO YOU KNOW**
>
> **The address space of IPv4 is 2^{32} or 4,294, 967,296.**

Example 2. Change the following IP addresses from dotted-decimal notation to binary notation.

(a) 111.56.45.78
(b) 221.34.7.82
(c) 241.8.56.12
(d) 75.45.34.78

Solution. We replace each decimal number with its binary equivalent :

(a) 01101111 00111000 00101101 01001110
(b) 11011101 00100010 00000111 01010010
(c) 11110001 00001000 00111000 00001100
(d) 01001011 00101101 00100010 01001110

Example 3. Find the error, if any, in the following IP addresses:

(a) 111.56.045.78
(b) 221.34.7.8.20
(c) 75.45.301.14
(d) 11100010.23.14.67

Solution.

(a) There are no leading zeroes in dotted-decimal notation (045).
(b) We may not have more than four numbers in an IP address.
(c) In dotted-decimal notation, each number is less than or equal to 255; 301 is outside this range.
(d) A mixture of binary notation and dotted-decimal notation is not allowed.

Example 4. Change the following IP addresses from binary notation to hexadecimal notation.

(a) 10000001 00001011 00001011 11101111
(b) 11000001 10000011 00011011 11111111

Solution. We replace each group of 4 bits with its hexadecimal equivalent (see Appendix B). Note that hexadecimal notation normally has no added spaces or dots; however, 0X (or 0x) is added at the beginning or the subscript 16 at the end to show that the number is in hexadecimal.

(a) 0X810B0BEF or $810B0BEF_{16}$
(b) 0XC1831BFF or $C1831BFF_{16}$

2.4. CLASSFUL ADDRESSING [MDU Dec. 2007; Dec. 2006]

IP addresses, when started a few decades ago, used the concept of classes. This architecture is called classful addressing. In the mid-1990s, a new architecture, called classless addressing, was introduced and will eventually supersede the original architecture. However, part of the Internet is still using classful addressing, but the migration is very fast.

Occupation of the Address Space

TABLE 2.1. Addresses Per Class

Class	Number of Addresses	Percentage
A	2^{31} = 2,147,483,648	50%
B	2^{30} = 1,073,741,824	25%
C	2^{29} = 536,870,912	12.5%
D	2^{28} = 268,435,456	6.25%
E	2^{28} = 268,435,456	6.25%

When IP addresses were first started they used the concept of classes. The range of IP addresses were divided into five classes:

1. Class A
2. Class B
3. Class C
4. Class D

Class A used up 50% of the address space, class B used up 25%, class C used up 12.5%, class D used up 6.25%, and class E also used up 6.25%.

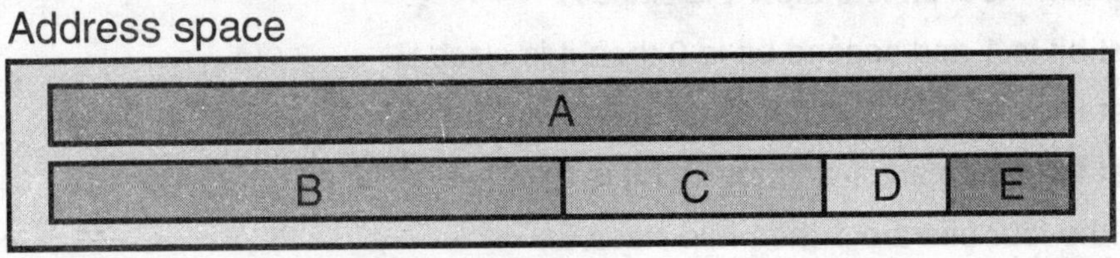

Fig. 2.22.

The way you recognize which class an IP address belongs to is by analyzing the first byte. If the number in the first byte is between 0-127, then the IP address is in the Class A range. If it is between 128-191 it is in Class B. If it is between 192-223 it is in the Class C range. If it is between 224-239 it is in the Class D range, and if it is between 240-255, then it belongs to Class E.

2.4.1. Class in Decimal Notations

	First byte	Second byte	Third byte	Fourth byte
Class A	0 to 127			
Class B	128 to 191			
Class C	192 to 223			
Class D	224 to 239			
Class E	240 to 255			

Fig. 2.23.

Example 1. If MDU IP address is 140.192.23.180. Looking at this address we can see that the first byte is 140. Since 140 is between the numbers 128-191, we know that it is in the Class B range.

Class in Binary Notation

	First byte	Second byte	Third byte	Fourth byte
Class A	0			
Class B	10			
Class C	110			
Class D	1110			
Class E	1111			

Fig. 2.24.

Finding the Address Class in Binary Notation

If first left most bit is 0 then it is class A

If first bit is 1 and second bit is 0 then it is class B

If first two bits are 1 and third bit is 0 then it is class C

If first three bits are 1 and fourth bit is 0 then it is class D

If all the four bits are 1 then it is class E

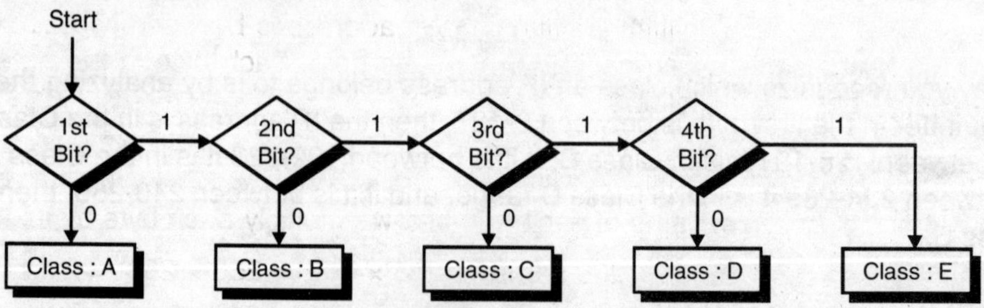

Fig. 2.25.

Example 2. How can you prove that we have 2,147,483,648 addresses in class A?

Solution. In class A, only 1 bit defines the class. The remaining 31 bits are available for the address. With 31 bits, we can have 2^{31} or 2,147,483,648 addresses.

Example 3. Find the class of each address:

(a) 00000001 00001011 00001011 11101111

(b) 11000001 10000011 00011011 11111111

(c) 10100111 11011011 10001011 01101111

(d) 11110011 10011011 11111011 00001111

Solution.

(a) The first bit is 0. This is a class A address.

(b) The first 2 bits are 1; the third bit is 0. This is a class C address.

(c) The first bit is 1; the second bit is 0. This is a class B address.

(d) The first 4 bits are 1s. This is a class E address

Example 4. Find the class of each address:

(a) 227.12.14.87

(b) 193.14.56.22

(c) 14.23.120.8

(d) 252.5.15.111

(e) 134.11.78.56

Solution.

(a) The first byte is 227 (between 224 and 239); the class is D.

(b) The first byte is 193 (between 192 and 223); the class is C.

(c) The first byte is 14 (between 0 and 127); the class is A.

(d) The first byte is 252 (between 240 and 255); the class is E.

(e) The first byte is 134 (between 128 and 191); the class is B.

Example 5. In Example 4 we showed that class A has 2^{31} (2,147,483,648) addresses. How can we prove this same fact using dotted-decimal notation?

Solution. The addresses in class A range from 0.0.0.0 to 127.255.255.255. We need to show that the difference between these two numbers is 2,147,483,648. This is a good exercise because it shows us how to define the range of addresses between two addresses. We notice that we are dealing with base 256 numbers here. Each byte in the notation has a weight. The weights are as follows :

$$256^3,\ 256^2,\ 256^1,\ 256^0$$

Now to find the integer value of each number, we multiply each byte by its weight:

Last address : $127 \times 256^3 + 255 \times 256^2 + 255 \times 256^1 + 255 \times 256^0 = 2,147,483,647$

First address : = 0

If we subtract the first from the last and add 1 to the result (remember we always add 1 to get the range), we get 2,147,483,648 or 2^{31}.

2.5. NETID AND HOSTID

An IP address is divided into

(1) Network ID (netid)

(2) Host ID (hostid)

The lengths of the netid vary depending on the class the IP address belongs to. In class A the netid occupies the first byte and the hostid occupies the remaining three bytes. In class B the netid occupies the first two bytes and the hostid occupies the remaining two bytes. In class C the first three bytes define the netid and the last remaining byte defines the hostid. Class D and E are not divided into netid and hostid. The following figure shows how the netid and hostid are divided in Classes A, B, and C.

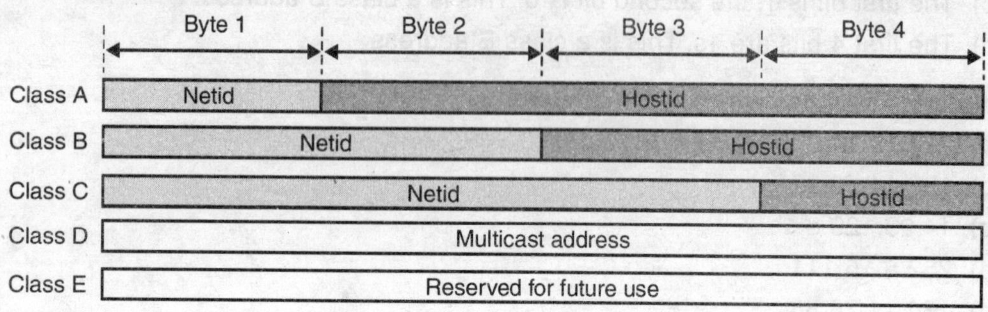

Fig. 2.26.

In classful addressing the netid and hostid are easily distinguishable by looking at the IP address. First you have to determine which class the IP address belongs to and from there you can tell which part is the netid and which part is the hostid. If it is in Class A, then the first byte represents the netid and the last three represent the hostid, and so on.

> **DO YOU KNOW**
>
> **Class D addresses are used for multicasting; there is only one block in this class and Class E addresses are reserved for future purposes; most of the block is wasted.**

Disadvantages of Class Full Addressing

It wastes a lot of IP addresses and since the Internet keeps growing larger, we can't afford to throw away IP addresses.

That is why a new addressing scheme was devised. It is called classless addressing because it doesn't use the classes which were used in classful addressing.

Example 1. Given the network address 17.0.0.0, find the class, the block, and the range of the addresses.

Solution. The class is A because the first byte is between 0 and 127. The block has a netid of 17. The addresses range from 17.0.0.0 to 17.255.255.255.

> **DO YOU KNOW**
>
> **Millions of class A and class B addresses and are wasted in class full addressing.**

Example 2. Given the network address 132.21.0.0, find the class, the block, and the range of the addresses.

Solution. The class is B because the first byte is between 128 and 191. The block has a netid of 132.21. The addresses range from 132.21.0.0 to 132.21.255.255.

DO YOU KNOW
The number of addresses in class C is smaller than the needs of most organizations.

Example 3. Given the network address 220.34.76.0, find the class, the block, and the range of the addresses.

Solution. Class is C because the first byte is between 192 and 223. The block has a netid of 220.34.76. The addresses range from 220.34.76.0 to 220.34.76.255.

2.6. CLASS LESS ADDRESSING

In classless addressing the entire range of IP addresses is divided into blocks of variable sizes. Depending on the needs of an organization, it is assigned an appropriate IP address along with a mask that denotes which part is the netid and which part is the hostid.

The format of a classless IP address is x.y.x.t /n.

Example 1. Given this classless IP address 140.120.80.0/16 provide the netid.

Solution. The /16 means that the first two bytes or 16 bits of the IP address 140.120.80.0 represent the netid. Since the first 16 bits represent the number 140.120. This means that the netid is 140.120.0.0.

In classless addressing variable-length blocks are assigned that belong to no class. In this architecture, the entire address space (232 addresses) is divided into blocks of different sizes.

Example 2. Which of the following can be the beginning address of a block that contains 16 addresses?

(a) 205.16.37.32
(b) 190.16.42.44
(c) 17.17.33.80
(d) 123.45.24.52

Solution. Only two are eligible (a and c). The address 205.16.37.32 is eligible because 32 is divisible by 16. The address 17.17.33.80 is eligible because 80 is divisible by 16.

Example 3. Which of the following can be the beginning address of a block that contains 256 addresses?

(a) 205.16.37.32
(b) 190.16.42.0
(c) 17.17.32.0
(d) 123.45.24.52

Solution. In this case, the right-most byte must be 0. As we mentioned in Chapter 4, the IP addresses use base 256 arithmetic. When the right-most byte is 0, the total address is divisible by 256. Only two addresses are eligible (b and c).

Example 4. Which of the following can be the beginning address of a block that contains 1024 addresses?

(a) 205.16.37.32
(b) 190.16.42.0
(c) 17.17.32.0
(d) 123.45.24.52

Solution. In this case, we need to check two bytes because 1024 = 4 × 256. The right-most byte must be divisible by 256. The second byte (from the right) must be divisible by 4. Only one address is eligible (c).

> **DO YOU KNOW**
> **The format of IP address in classless adressing is x.y.x.t/n.**

TABLE 2.2. Prefix lengths

/n	Mask	/n	Mask	/n	Mask	/n	Mask
/1	128.0.0.0	/9	255.128.0.0	/17	255.255.128.0	/26	255.255.255.192
/2	192.0.0.0	/10	255.192.0.0	/18	255.255.192.0	/26	255.255.255.192
/3	224.0.0.0	/11	255.224.0.0	/19	255.255.224.0	/27	255.255.255.224
/4	240.0.0.0	/12	255.240.0.0	/20	255.255.240.0	/28	255.255.255.240
/5	248.0.0.0	/13	255.248.0.0	/21	255.255.248.0	/29	255.255.255.248
/6	252.0.0.0	/14	255.252.0.0	/22	255.255.252.0	/30	255.255.255.252
/7	254.0.0.0	/15	255.254.0.0	/23	255.0.255.254.0	/31	255.255.255.254
/8	255.0.0.0	/16	255.255.0.0	/24	255.255.255.0	/32	255.255.255.255

Example 5. What is the first address in the block if one of the addresses is 167.199.170.82/27?

Solution. The prefix length is 27, which means that we must keep the first 27 bits as is and change the remaining bits (5) to 0s.

The following shows the process :

Address in binary: 10100111 11000111 10101010 01010010

Keep the left 27 bits: 10100111 11000111 10101010 01000000

Result in CIDR notation: 167.199.170.64/27

Example 6. What is the first address in the block if one of the addresses is 140.120.84.24/20?

Solution. Following Figure 2.27 shows the solution. The first, second, and fourth bytes are easy; for the third byte we keep the bits corresponding to the number of 1s in that group. The first address is 140.120.80.0/20.

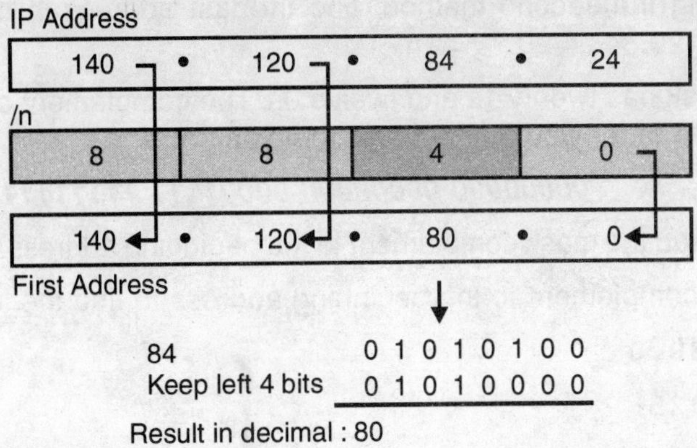

Fig. 2.27.

Example 7. Find the first address in the block if one of the addresses is 140.120.84.24/20.

Solution. The first, second, and fourth bytes are as defined in the previous example. To find the third byte, we write 84 as the sum of powers of 2 and select only the leftmost 4 (m is 4) as shown in Figure below. The first address is 140.120.80.0/20.

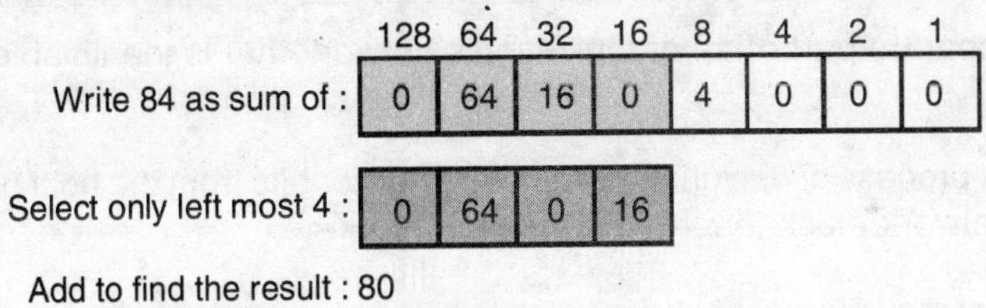

Fig. 2.28.

Example 8. Find the number of addresses in the block if one of the addresses is 140.120.84.24/20.

Solution. The prefix length is 20. The number of addresses in the block is 2^{32-20} or 2^{12} or 4096. Note that this is a large block with 4096 addresses.

Example 9. Using the first method, find the last address in the block if one of the addresses is 140.120.84.24/20.

Solution. We found in the previous examples that the first address is 140.120.80.0/20 and the number of addresses is 4096. To find the last address, we need to add 4095 (4096 – 1) to the first address.

To keep the format in dotted-decimal notation, you need to represent 4095 in base 256 and do the calculation in base 256. We write 4095 as 15.255. We then add the first address to this number (in base 255) to obtain the last address as shown below:

$$\begin{array}{r} 140\,.\,120\,.\,80\,.\,0 \\ 15\,.\,255 \\ \hline 140\,.\,120\,.\,95\,.\,255 \end{array}$$

The last address is 140.120.95.255/20.

Example 10. Using the second method, find the last address in the block if one of the addresses is 140.120.84.24/20.

Solution. The mask has twenty 1s and twelve 0s. The complement of the mask has twenty 0s and twelve 1s. In other words, the mask complement is

00000000 00000000 00001111 11111111

or 0.0.15.255. We add the mask complement to the beginning address to find the last address

We add the mask complement to the beginning address to find the last address.

```
14.120.80.0
0.0.15.225
....................................
140.120.95.255
....................................
The last address is 140.120.95.255/20.
```

2.7. SUNETTING [MDU May 2008]

A subnet is a separate part of an organization's network that is identifiable.

OR

Subnetting is the process of designating some high-order bits from the host part and grouping them with the network mask to form the **subnet mask**.

Typically, a subnet may represent all the machines at one geographic location, in one building, or on the same local area network (LAN). Having an organization's network divided into subnets allows it to be connected to the Internet with a single shared network address. Without subnets, an organization could get multiple connections to the Internet, one for each of its physically separate subnetworks, but this would require an unnecessary use of the limited number of network numbers the Internet has to assign. It would also require that Internet routing tables on gateways outside the organization would need to know about and have to manage routing that could and should be handled within an organization.

> **DO YOU KNOW**
>
> **When an organization is granted a block of addresses, it can create subnets to meet its needs. The prefix length increases to define the subnet prefix length.**

The Internet is a collection of networks whose users communicate with each other. Each communication carries the address of the source and destination networks and the particular machine within the network associated with the user or host computer at each end. This address is called the IP address (Internet Protocol address).

This 32-bit IP address has two parts :

1. One part identifies the network (with the *network number*) and the
2. Second part identifies the specific machine or host within the network (with the *host number*).

An organization can use some of the bits in the machine or host part of the address to identify a specific subnet. Effectively, the IP address then contains three parts: the network number, the subnet number, and the machine number.

2.7.1. Subnet Mask

The subnet mask not only determines the size of a subnet, but it can also help you pinpoint where the end points on the subnet are if you're given any IP address within that subnet. The reason it's called a subnet "mask" is that it literally masks out the host bits and leaves only the Network ID that begins the subnet. Once you know the beginning of the subnet and how big it is, you can determine the end of the subnet, which is the Broadcast ID.

Applying a subnet mask to an IP address allows you to identify the network and node parts of the address. The network bits are represented by the 1s in the mask, and the node bits are represented by the 0s. Performing a bitwise logical AND operation between the IP address and the subnet mask results in the *Network Address* or Number. For example, using our test IP address and the default Class B subnet mask, we get :

```
Binary                                 Decimal

10001100.10110011.11110000.11001000    140.179.240.200 Class B IP Address

11111111.11111111.00000000.00000000    255.255.000.000 Default Class B Subnet Mask
_____________________________________

10001100.10110011.00000000.00000000    140.179.000.000   Network Address
```

TIPS

Masking extracts network address from IP address.

2.7.2. Default Subnet Masks

- **Class A** - 255.0.0.0 - 11111111.00000000.00000000.00000000
- **Class B** - 255.255.0.0 - 11111111.11111111.00000000.00000000
- **Class C** - 255.255.255.0 - 11111111.11111111.11111111.00000000

Sunet mask is of two types :

1. Default mask
2. Subnet mask

DO YOU KNOW

In order to define the range of IP addresses in subnet-ting we need the :

The first address of the subnet + subnet mask

Example 1. An organization is granted the block 130.34.12.64/26. The organization needs 4 subnets. What is the subnet prefix length?

Solution. We need 4 subnets, which means we need to add two more 1s ($\log_2 4 = 2$) to the site prefix. The subnet prefix is then /28.

Example 2. What are the subnet addresses and the range of addresses for each subnet in the 130.34.12.64/26?

Solution. The site has $2^{32-26} = 64$ addresses. Each subnet has $2^{32-28} = 16$ addresses. Now let us find the first and last address in each subnet.

1. The first address in the first subnet is 130.34.12.64/28, using the procedure we showed in the previous examples. Note that the first address of the first subnet is the first address of the block. The last address of the subnet can be found by adding 15 (16–1) to the first address. The last address is 130.34.12.79/28.

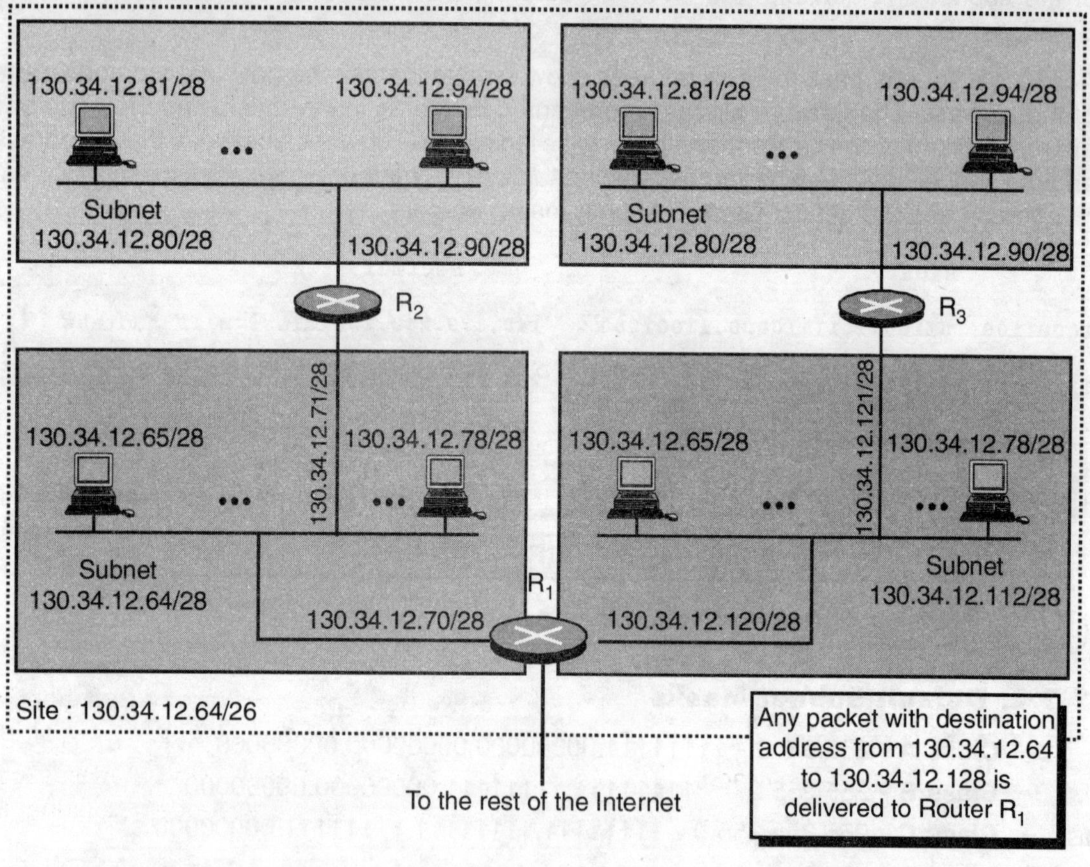

Fig. 2.29.

2. The first address in the second subnet is 130.34.12.80/28; it is found by adding 1 to the last address of the previous subnet. Again adding 15 to the first address, we obtain the last address, 130.34.12.95/28.
3. Similarly, we find the first address of the third subnet to be 130.34.12.96/28 and the last to be 130.34.12.111/28.
4. Similarly, we find the first address of the fourth subnet to be 130.34.12.112/28 and the last to be 130.34.12.127/28.

Example 3. An organization is granted a block of addresses with the beginning address 14.24.74.0/24. There are $2^{32-24} = 256$ addresses in this block. The organization needs to have 11 subnets as shown below:

(a) Two subnets, each with 64 addresses.

(b) Two subnets, each with 32 addresses.

(c) Three subnets, each with 16 addresses.

(d) Four subnets, each with 4 addresses.

Design the subnets.

Solution.

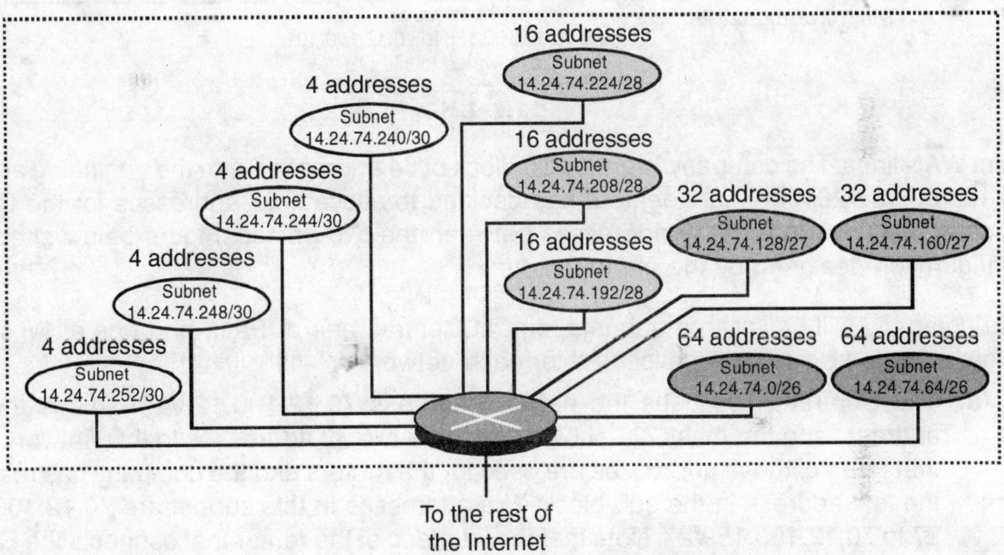

Fig. 2.30.

1. We use the first 128 addresses for the first two subnets, each with 64 addresses. Note that the mask for each network is /26. The subnet address for each subnet is given in the figure.
2. We use the next 64 addresses for the next two subnets, each with 32 addresses. Note that the mask for each network is /27. The subnet address for each subnet is given in the figure.
3. We use the next 48 addresses for the next three subnets, each with 16 addresses. Note that the mask for each network is /28. The subnet address for each subnet is given in the figure.
4. We use the last 16 addresses for the last four subnets, each with 4 addresses. Note that the mask for each network is /30. The subnet address for each subnet is given in the figure.

Example 4. Assume a company has three offices: Central, East, and West. The Central office is connected to the East and West offices via private, point-to-

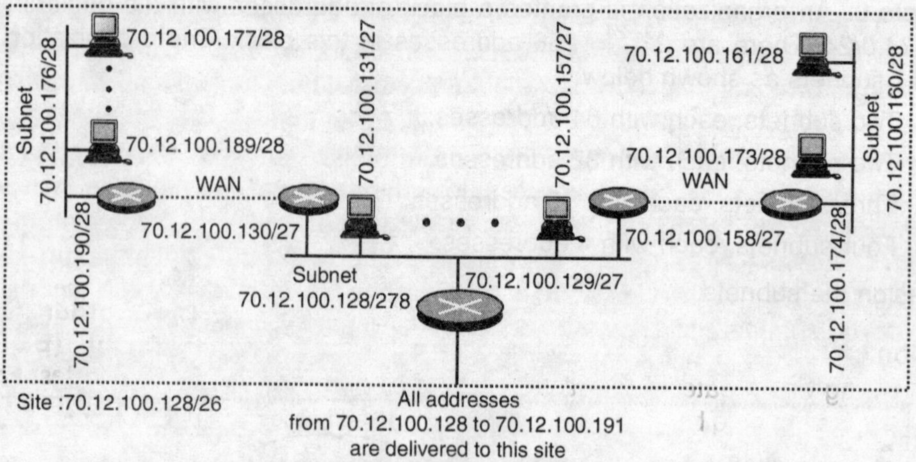

Fig. 2.31.

point WAN lines. The company is granted a block of 64 addresses with the beginning address 70.12.100.128/26. The management has decided to allocate 32 addresses for the Central office and divides the rest of addresses between the two offices. Figure below shows the configuration designed by the management.

The company will have three subnets, one at Central, one at East, and one at West. The following lists the subblocks allocated for each network :

(a) The Central office uses the network address 70.12.100.128/27. This is the first address, and the mask /27 shows that there are 32 addresses in this network. Note that three of these addresses are used for the routers and the company has reserved the last address in the sub-block. The addresses in this subnet are 70.12.100.128/27 to 70.12.100.159/27. Note that the interface of the router that connects the Central subnet to the WAN needs no address because it is a point-to-point connection.

(b) The West office uses the network address 70.12.100.160/28. The mask /28 shows that there are only 16 addresses in this network. Note that one of these addresses is used for the router and the company has reserved the last address in the sub-block. The addresses in this subnet are 70.12.100.160/28 to 70.12.100.175/28. Note also that the interface of the router that connects the West subnet to the WAN needs no address because it is a point-to- point connection.

(c) The East office uses the network address 70.12.100.176/28. The mask /28 shows that there are only 16 addresses in this network. Note that one of these addresses is used for the router and the company has reserved the last address in the sub-block. The addresses in. this subnet are 70.12.100.176/28 to 70.12.100.191/28. Note also that the interface of the router that connects the East subnet to the WAN needs no address because it is a point-to-point connection.

2.8. SUPERNETTING

[MDU May 2008; Dec. 2006]

Supernetting, also called Classless Inter-Domain Routing (CIDR) is a way to aggregate multiple internet addresses of the same class. The original internet protocol (IP) defines IP addresses in four . Major classes of address structure, class A through D. Each class allocates one portion to 32-bit internet address format to a network address and the remaining portion of the specific host machines within the network.

Using supernetting the network address 192.168.2.0/24 and adjacent address 192.168.3.0/24 can be merged into 192.168.2.0/23. The "23" at the end of the address says that the first 23 bits are the network part of the address, leaving the remaining nine bits for specific host addresses.

Supernetting is most often used to combine class C network addresses and is the basis for most routing protocols currently used on the internet.

Supernetting was created as a way to solve the problem of routing tables growing beyond the ability of current software and people to manage and to provide a solution to the exhaustion of class B network address space.

DO YOU KNOW

The Border Gateway Protocol (BGP) and Open Shortest Path First (OSPF) both support supernetting.

Rules for Supernetting

- ➢ The number of blocks must be a power of 2.
- ➢ The blocks must be contiguous in the address space (no gaps between the blocks).
- ➢ The third byte of the first address in the superblock must be evenly divisible by the number of blocks.

DO YOU KNOW

In order to define the range of IP addresses In super-netting we need The first address of the supernet + supernet mask.

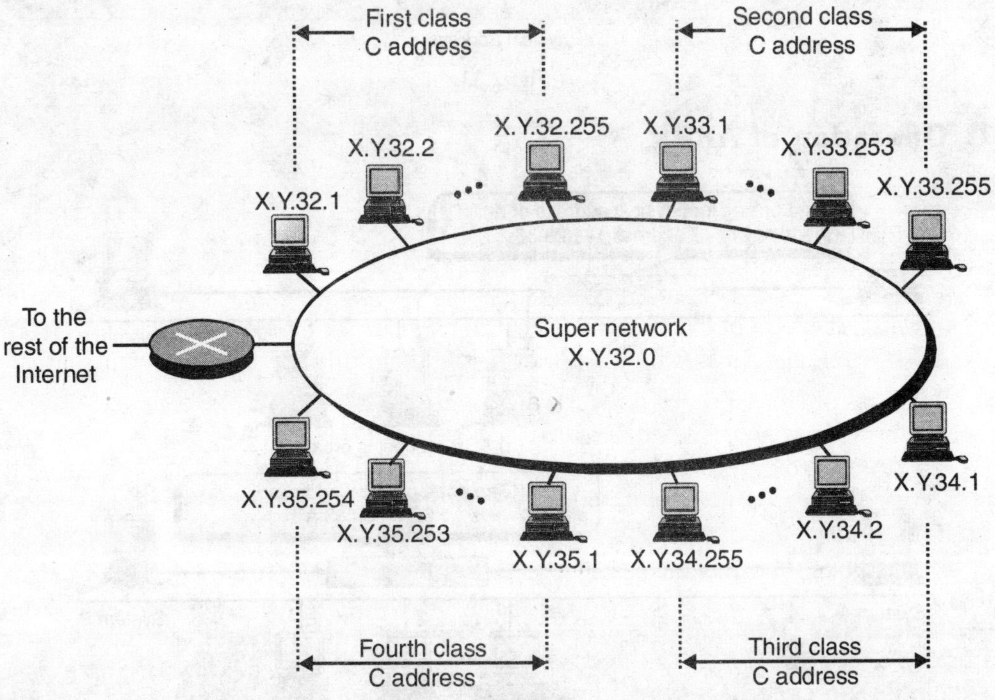

Fig. 2.32.

Example 1. A company needs 1000 addresses. Which of the following set of class C blocks can be used to form a supernet for this company?

(a) 198.47.32.0 198.47.33.0 198.47.34.0

(b) 198.47.32.0 198.47.42.0 198.47.52.0 198.47.62.

(c) 198.47.31.0 198.47.32.0 198.47.33.0 198.47.52.

(d) 198.47.32.0 198.47.33.0 198.47.34.0 198.47.35.

Solution. Only d can be used to form the subnet. In a there are three blocks and for supernetting four blocks are required, in b blocks are not contiguous. In c third byte of first block is not divisible by 4.ke a supernetwork out of 16 class C blocks. What is thet mask?

2.9. ADDRESS RESOLUTION PROTOCOL (ARP) [MDU Dec. 2008; May 2008]

It converts logical address to physical address. ARP associates an IP address with its physical address. On a typical physical network, such as a LAN, each device on a link is identified by a physical or station address that is usually imprinted on the NIC.

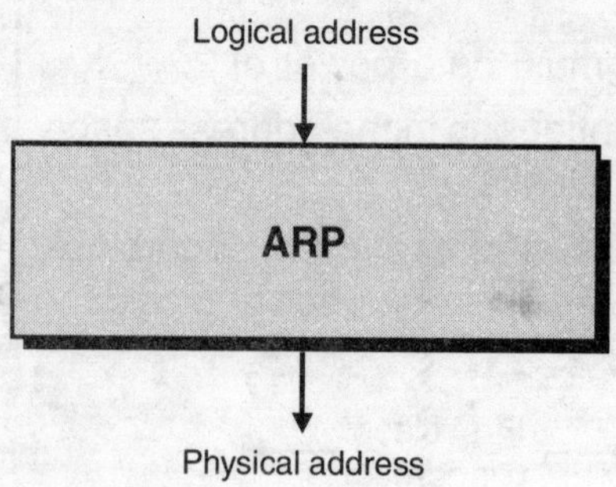

Fig. 2.33.

2.9.1. Operation of ARP

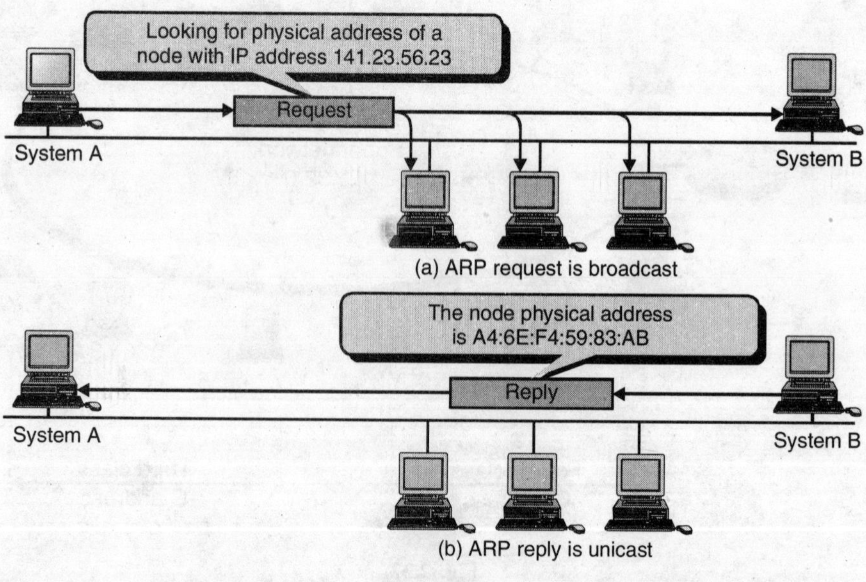

Fig. 2.34.

2.9.2. Encapsulation of ARP Packet

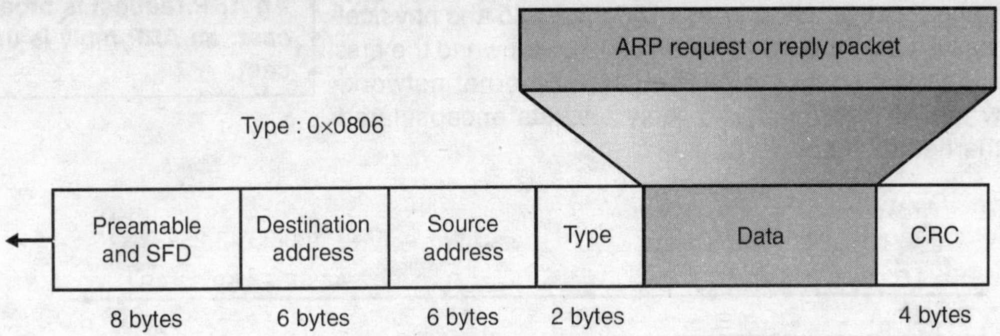

Fig. 2.35.

There are four possible case using ARP :

CASE 1: A host has a packet to send another hoston the same network

CASE 2 : A host wants to send a packet to another host on another network

CASE 3 : A router receives a packet to be sent to host on another network.It must first be delivered to aapropriate router

CASE 4 : A router receives a packet to be sent to a host on the same network.

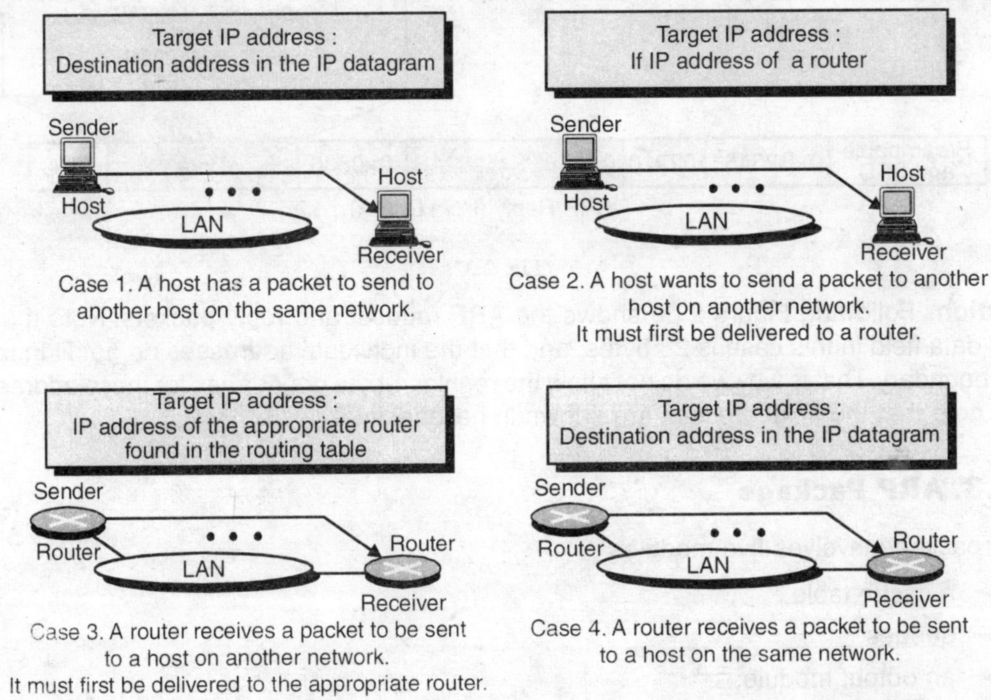

Fig. 2.36.

> **DO YOU KNOW**
>
> **An ARP request is broadcast; an ARP reply is unicast.**

Example 1. A host with IP address 130.23.43.20 and physical address B2:34:55:10:22:10 has a packet to send to another host with IP address 130.23.43.25 and physical address A4:6E:F4:59:83:AB (which is unknown to the first host). The two hosts are on the same Ethernet network. Show the ARP request and reply packets encapsulated in Ethernet frames.

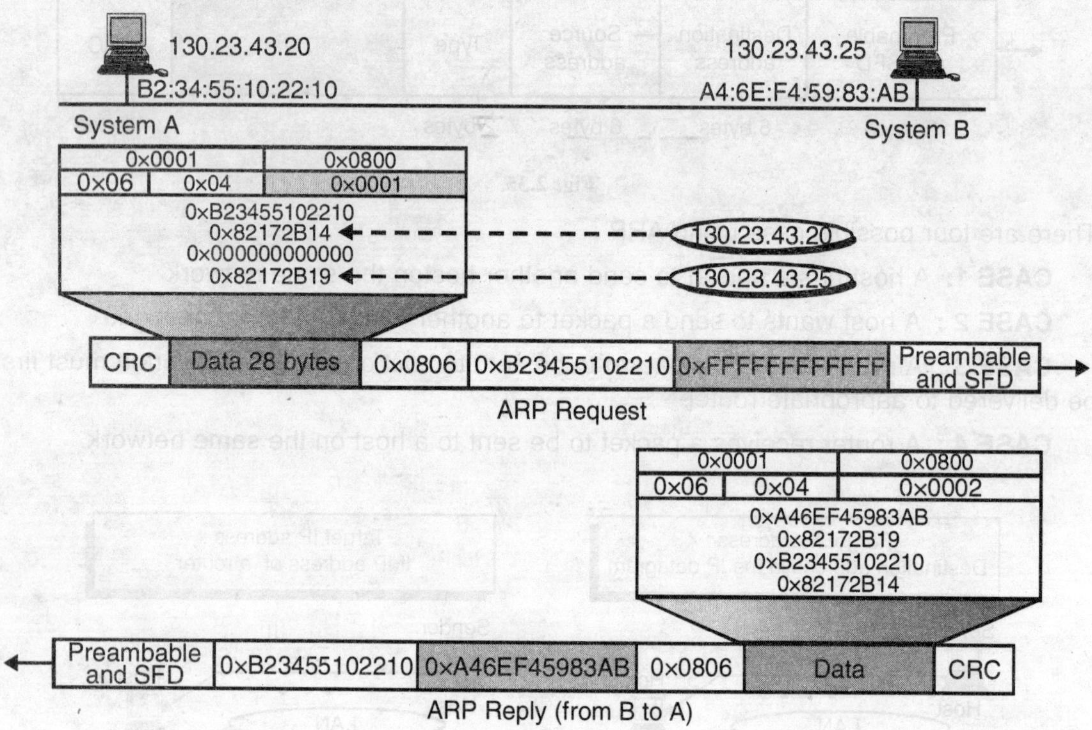

Fig. 2.37.

Solution. Following Figure 2.39 shows the ARP request and reply packets. Note that the ARP data field in this case is 28 bytes, and that the individual addresses do not fit in the 4-byte boundary. That is why we do not show the regular 4-byte boundaries for these addresses. Also note that the IP addresses are shown in hexadecimal.

2.9.3. ARP Package

ARP package involves five modules :

- ➢ a cache table,
- ➢ queues,
- ➢ an output module,
- ➢ an input module,
- ➢ and a cache-control module

2.9.4. ARP Components

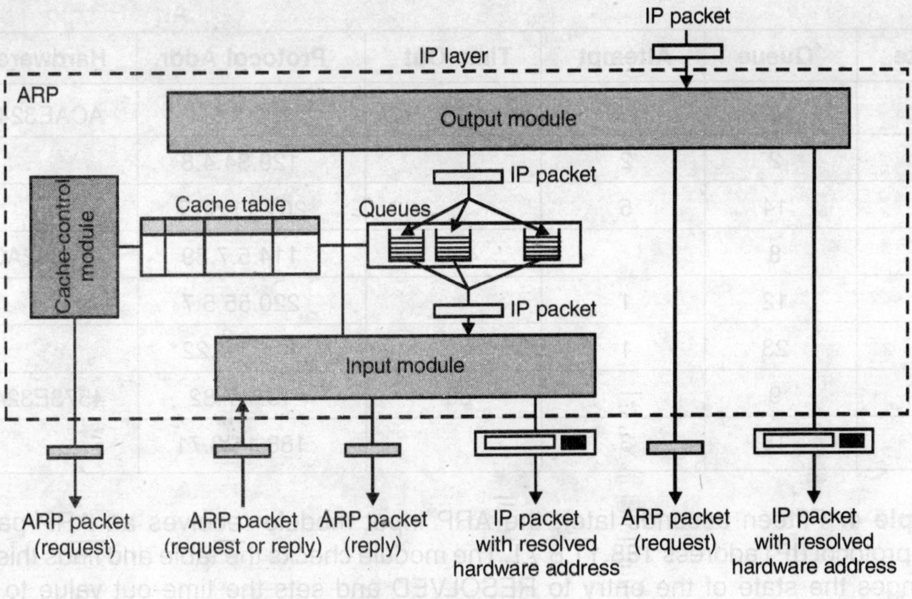

Fig. 2.38.

Original Cache Table Used for Examples

State	Queue	Attempt	Time-Out	Protocol Addr.	Hardware Addr.
R	5		900	180.3.6.1	ACAE32457342
P	2	2		129.34.4.8	
P	14	5		201.11.56.7	
R	8		450	114.5.7.89	457342ACAE32
P	12	1		220.55.5.7	
F					
R	9		60	19.1.7.82	4573E3242ACA
P	18	3		188.11.8.71	

Example 2. The ARP output module receives an IP datagram (from the IP layer) with the destination address 114.5.7.89. It checks the cache table and finds that an entry exists for this destination with the RESOLVED state (R in the table). It extracts the hardware address, which is 457342ACAE32, and sends the packet and the address to the data link layer for transmission. The cache table remains the same.

Example 3. Twenty seconds later, the ARP output module receives an IP datagram (from the IP layer) with the destination address 116.1.7.22. It checks the cache table and does not find this destination in the table. The module adds an entry to the table with the state PENDING and the Attempt value 1. It creates a new queue for this destination and enqueues the packet. It then sends an ARP request to the data link layer for this destination. The new cache table is shown in following Table 7.2.

Updated cache table for Example 3.

State	Queue	Attempt	Time-Out	Protocol Addr.	Hardware Addr.
R	5		900	180.3.6.1	ACAE32457342
P	2	2		129.34.4.8	
P	14	5		201.11.56.7	
R	8		450	114.5.7.89	457342ACAE32
P	12	1		220.55.5.7	
P	23	1		116.1.7.22	
R	9		60	19.1.7.82	4573E3242ACA
P	18	3		188.11.8.71	

Example 4. Fifteen seconds later, the ARP input module receives an ARP packet with target protocol (IP) address 188.11.8.71. The module checks the table and finds this address. It changes the state of the entry to RESOLVED and sets the time-out value to 900. The module then adds the target hardware address (E34573242ACA) to the entry. Now it accesses queue 18 and sends all the packets in this queue, one by one, to the data link layer. The new cache table is shown in foolowing table.

Updated cache table for Example 4

State	Queue	Attempt	Time-Out	Protocol Addr.	Hardware Addr.
R	5		900	180.3.6.1	ACAE32457342
P	2	2		129.34.4.8	
P	14	5		201.11.56.7	
R	8		450	114.5.7.89	457342ACAE32
P	12	1		220.55.5.7	
P	23	1		116.1.7.22	
R	9		60	19.1.7.82	4573E3242ACA
P	18		900	188.11.8.71	E34573242ACA

Example 5. Twenty-five seconds later, the cache-control module updates every entry. The time-out values for the first three resolved entries are decremented by 60. The time-out value for the last resolved entry is decremented by 25. The state of the next-to-the last entry is changed to FREE because the time-out is zero. For each of the three pending entries, the value of the attempts

Updated cache table for Example 5

State	Queue	Attempt	Time-Out	Protocol Addr.	Hardware Addr.
R	5		840	180.3.6.1	ACAE32457342
P	2	3		129.34.4.8	
F					
R	8		390	114.5.7.89	457342ACAE32
P	12	2		220.55.5.7	
P	23	2		116.1.7.22	
F					
R	18		875	188.11.8.71	E34573242ACA

2.10. REVERSE ADDRESS RESOLUTION PROTOCOL (RARP)

[MDU Dec. 2008; Dec. 2006]

It converts logical address to physical address.

OR

RARP finds the logical address for a machine that only knows its physical address.

> **DO YOU KNOW**
>
> **The RARP request packets are broadcast; the RARP reply packets are unicast.**

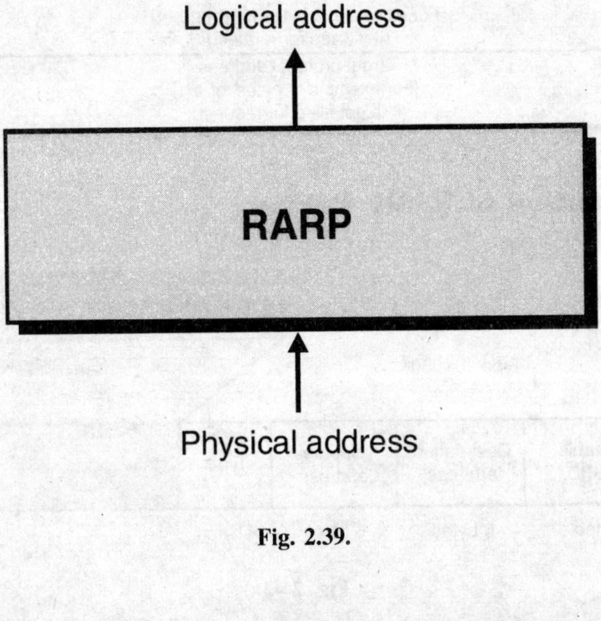

Fig. 2.39.

2.10.1. RARP Operation

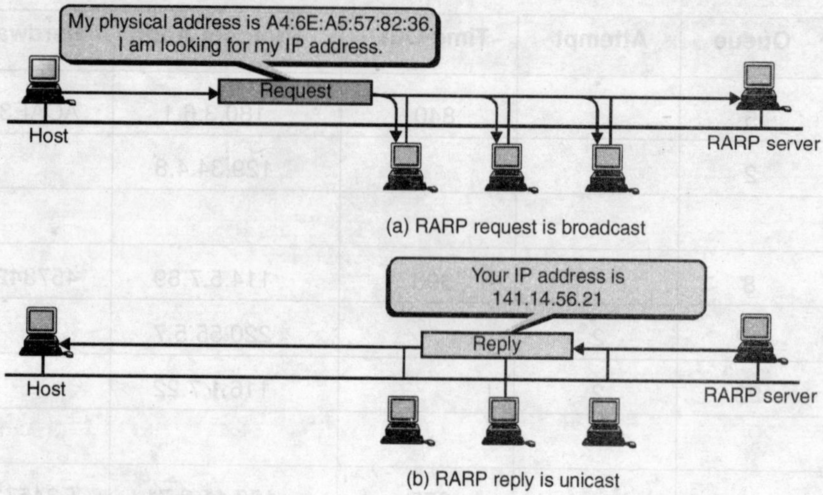

Fig. 2.40.

2.10.2. RARP Packet Format

Hardware type		Protocol type
Hardware length	Protocol length	Operation Request 3, Reply 4
Sender hardware address (For example, 6 bytes for Ethernet)		
Sender protocol address (For example, 4 bytes for IP) (It is not filled for request)		
Target hardware address (For example, 6 bytes for Ethernet) (It is filled for request)		
Target protocol address (For example, 4 bytes for IP) (It is not filled for request)		

Fig. 2.41.

2.10.3. Encapsulation of RARP Packet

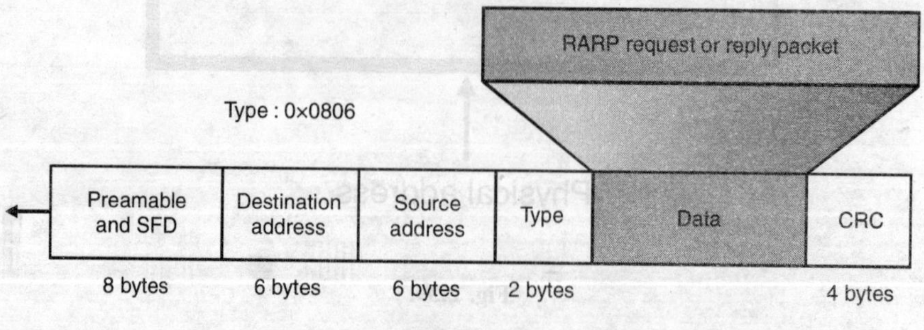

Fig. 2.42.

2.11. POSITION OF ARP AND RARP IN TCP/IP PROTOCOL SUITE

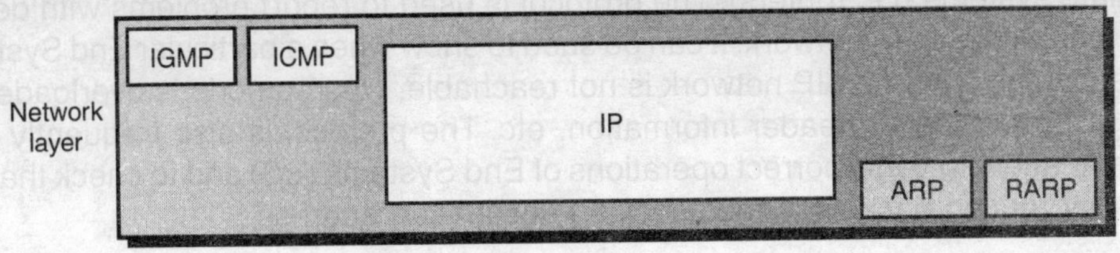

Fig. 2.43.

2.12. ICMP [MDU Dec. 2008; May 2008; Dec. 2006]

The Internet Control Message Protocol (ICMP) is a helper protocol that supports IP with facility for Error reporting Simple queries.

ICMP messages are sent in several situations: for example,

- ➢ when a datagram cannot reach its destination,
- ➢ when the gateway does not have the buffering capacity to forward a datagram,
- ➢ when the gateway can direct the host to send traffic on a shorter route.

The Internet Protocol is not designed to be absolutely reliable. The purpose of these control messages is to provide feedback about problems in the communication environment, not to make IP reliable. There are still no guarantees that a datagram will be delivered or a control message will be returned. Some datagrams may still be undelivered without any report of their loss. The higher level protocols that use IP must implement their own reliability procedures if reliable communication is required. The ICMP messages typically report errors in the processing of datagrams. To avoid the infinite regress of messages about messages etc., no ICMP messages are sent about ICMP messages.

DO YOU KNOW

ICMP is considered an integral part of IP.

TIPS

ICMP provides error reporting, flow control and first-hop gateway redirection.

2.12.1. Position of ICMP in the Network Layer

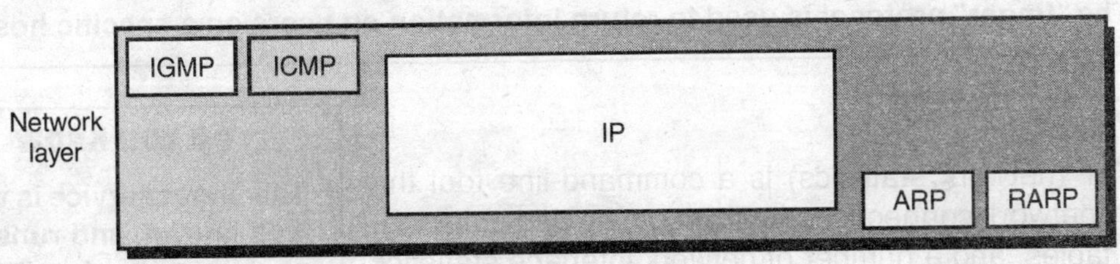

Fig. 2.44.

The Internet Control Message Protocol (ICMP) protocol is classic example of a client server application. The ICMP server executes on all IP end system computers and all IP intermediate systems (i.e. routers). The protocol is used to report problems with delivery of IP datagrams within an IP network. It can be sued to show when a particular End System (ES) is not responding, when an IP network is not reachable, when a node is overloaded, when an error occurs in the IP header information, etc. The protocol is also frequently used by Internet managers to verify correct operations of End Systems (ES) and to check that routers are correctly routing packets to the specified destination address

2.12.2. Types of Messages

ICMP messages are divided into error-reporting messages and query messages. The error-reporting messages report problems that a router or a host (destination) may encounter. The query messages get specific information from a router or another host.

Fig. 2.45.

2.13. ICMP MESSAGES

2.13.1. Finger Protocol

The Name/Finger protocol, is based on Request for comments document. I act as an interface to the **name** and **finger** programs that provide status reports on a particular computer system or a particular person at network sites

Finger is based on the Transmission Control Protocol, using TCP port 79 decimal. The local host opens a TCP connection to a remote host on the Finger port. An RUIP (Remote User Information Program) becomes available on the remote end of the connection to process the request. The local host sends the RUIP a one line query based upon the Finger query specification, and waits for the RUIP to respond. The RUIP receives and processes the query, returns an answer, then initiates the close of the connection. The local host receives the answer and the close signal, then proceeds closing its end of the connection.

TIPS

The "finger" protocol is used to return information on users on a specific host.

2.13.2. NETSTAT

NETSTAT (network statistics) is a command-line tool that displays network connections (both incoming and outgoing), routing tables, and a number of network interface statistics. It is available on Unix, Unix-like, and Windows NT-based operating systems.

DO YOU KNOW

The finger service is very well known, and runs on port seventy-nine (79).

Functions of NETSTAT

- It Displays active TCP connections,
- It displays ports on which the computer is listening,
- It displays Ethernet statistics,
- It shows the IP routing table,
- It shows IPv4 statistics (for the IP, ICMP, TCP, and UDP protocols),
- It shows IPv6 statistics (for the IPv6, ICMPv6, TCP over IPv6, and UDP over IPv6 protocols).
- If netstat is used without parameters, **it** displays active TCP connections.

2.13.3. NETSTAT in Windows

Syntax

```
netstat [-a] [-e] [-n] [-o] [-p Protocol] [-r] [-s] [Interval]
```

where a, e,n,o,p,r,s are parameters having following functinalities :

-a : Displays all active TCP connections and the TCP and UDP ports on which the computer is listening.

-e : Displays Ethernet statistics, such as the number of bytes and packets sent and received. This parameter can be combined with **-s**.

-n : Displays active TCP connections, however, addresses and port numbers are expressed numerically and no attempt is made to determine names.

-o : Displays active TCP connections and includes the process ID (PID) for each connection. You can find the application based on the PID on the **Processes** tab in Windows Task Manager. This parameter can be combined with **-a**, **-n**, and **-p**.

-p *Protocol* : Shows connections for the protocol specified by *Protocol*. In this case, the *Protocol* can be **tcp**, **udp**, **tcpv6**, or **udpv6**. If this parameter is used with **-s** to display statistics by protocol, *Protocol* can be **tcp**, **udp**, **icmp**, **ip**, **tcpv6**, **udpv6**, **icmpv6**, or **ipv6**.

-s : Displays statistics by protocol. By default, statistics are shown for the TCP, UDP, ICMP, and IP protocols. If the IPv6 protocol for Windows XP is installed, statistics are shown for the TCP over IPv6, UDP over IPv6, ICMPv6, and IPv6 protocols. The **-p** parameter can be used to specify a set of protocols.

-r : Displays the contents of the IP routing table. This is equivalent to the **route print** command.

***Interval* :** Redisplays the selected information every *Interval* seconds. Press CTRL+C to stop the redisplay. If this parameter is omitted, **netstat** prints the selected information only once.

/? : Displays help at the command prompt.

TIPS

Parameters used with this command must be prefixed with a hyphen (-) rather than a slash (/).

NETSTAT provides statistics for the following :

1. **Proto :** The name of the protocol (TCP or UDP).
2. **Local Address :** The IP address of the local computer and the port number being used. The name of the local computer that corresponds to the IP address and the name of the port is shown unless the **-n** parameter is specified. If the port is not yet established, the port number is shown as an asterisk (*).
3. **Foreign Address :** The IP address and port number of the remote computer to which the socket is connected. The names that corresponds to the IP address and the port are shown unless the **-n** parameter is specified. If the port is not yet established, the port number is shown as an asterisk (*)
4. **(State)** - Indicates the state of a TCP connection. The possible states are as follows:
   - ➢ CLOSE_WAIT
   - ➢ CLOSED
   - ➢ ESTABLISHED
   - ➢ FIN_WAIT_1
   - ➢ FIN_WAIT_2
   - ➢ LAST_ACK
   - ➢ LISTEN
   - ➢ SYN_RECEIVED
   - ➢ SYN_SEND
   - ➢ TIMED_WAIT
5. This command is available only if the **Internet Protocol (TCP/IP)** protocol is installed as a component in the properties of a network adapter in Network Connections

Example 1. Write command To display both the Ethernet statistics and the statistics for all protocols.

Solution. netstat -e -s

Example 2. Write command To display the statistics for only the TCP and UDP protocols.

Solution. netstat -s -p tcp udp

Example 3. Write command To display active TCP connections and the process IDs every 5 seconds.

Solution. **nbtstat -o 5**

Example 4. Write command to display active TCP connections and the process IDs using numerical form.

Solution. nbtstat -n –o

Formatting legend

Format	Meaning
Italic	Information that the user must supply
Bold	Elements that the user must type exactly as shown
Ellipsis (...)	Parameter that can be repeated several times in a command line
Between brackets ([])	Optional items
Between braces ({}); choices separated by pipe (l). Example: {evenlodd}	Set of choices from which the user must choose only one
Courier font	Code or program output

2.13.4. NETSTAT in Linux

netstat is a useful tool for checking network configuration and activity. It is a collection of several tools combined together.

1. Displaying the Routing Table

When you invoke **netstat** with the –r flag, it displays the kernel routing table in the way we've been doing with **route**. On vstout, it produces :

```
# netstat -nr
Kernel IP routing table
Destination   Gateway      Genmask          Flags  MSS Window
irtt Iface
127.0.0.1     *  255.255.255.255 UH     0 0     0 lo
172.16.1.0    *  255.255.255.0   U      0 0     0 eth0
172.16.2.0    172.16.1.1   255.255.255.0    UG     0 0     0 eth0
```

The –n option makes **netstat** print addresses as dotted quad IP numbers rather than the symbolic host and network names. This option is especially useful when you want to avoid address lookups over the network (e.g., to a DNS or NIS server).

The second column of **netstat**?'s output shows the gateway to which the routing entry points. If no gateway is used, an asterisk is printed instead. The third column shows the "generality" of the route, i.e., the network mask for this route. When given an IP address to find a suitable route for, the kernel steps through each of the routing table entries, taking the bitwise AND of the address and the genmask before comparing it to the target of the route.

The fourth column displays the following flags that describe the route :

- **G** - The route uses a gateway.
- **U** - The interface to be used is up.
- **H** - Only a single host can be reached through the route. For example, this is the case for the loopback entry 127.0.0.1.
- **D** - This route is dynamically created. It is set if the table entry has been generated by a routing daemon like **gated** or by an ICMP redirect message
- **M** - This route is set if the table entry was modified by an ICMP redirect message.
- **!** - The route is a reject route and datagrams will be dropped.

The next three columns show the MSS, Window and irtt that will be applied to TCP connections established via this route.

The MSS is the Maximum Segment Size and is the size of the largest datagram the kernel will construct for transmission via this route.

The Window is the maximum amount of data the system will accept in a single burst from a remote host.

The acronym irtt stands for "initial round trip time." The TCP protocol ensures that data is reliably delivered between hosts by retransmitting a datagram if it has been lost. The TCP protocol keeps a running count of how long it takes for a datagram to be delivered to the remote end, and an acknowledgement to be received so that it knows how long to wait before assuming a datagram needs to retransmitted; this process is called the round-trip time. The initial round-trip time is the value that the TCP protocol will use when a connection is first established. For most network types, the default value is okay, but for some slow networks, notably certain types of amateur packet radio networks, the time is too short and causes unnecessary retransmission. The irtt value can be set using the **route** command. Values of zero in these fields mean that the default is being used.

Finally, the last field displays the network interface that this route will use.

2. Displaying Interface Statistics

When invoked with the –i flag, **netstat** displays statistics for the network interfaces currently configured. If the –a option is also given, it prints *all* interfaces present in the kernel, not only those that have been configured currently. On vstout, the output from **netstat** will look like this:

```
# netstat -i
```

Kernel Interface Table

```
    Iface MTU Met  RX-OK RX-ERR RX-DRP RX-OVR  TX-OK TX-ERR TX-
DRP TX-OVR Flags
       lo  0    0   3185     0      0      0   3185     0     0     0 BLRU
    eth0 1500    0 972633    17     20    120 628711   217     0     0 BRU
```

The MTU and Met fields show the current MTU and metric values for that interface. The RX and TX columns show how many packets have been received or transmitted error-free (RX-OK/TX-OK) or damaged (RX-ERR/TX-ERR); how many were dropped (RX-DRP/TX-DRP); and how many were lost because of an overrun (RX-OVR/TX-OVR).

The last column shows the flags that have been set for this interface. These characters are one-character versions of the long flag names that are printed when you display the interface configuration with **ifconfig**:

- B - A broadcast address has been set.
- L - This interface is a loopback device.
- M - All packets are received (promiscuous mode).
- O - ARP is turned off for this interface.
- P - This is a point-to-point connection.
- R - Interface is running.
- U - Interface is up.

3. Displaying Connections

netstat supports a set of options to display active or passive sockets. The options –t, –u, –w, and –x show active TCP, UDP, RAW, or Unix socket connections. If you provide the –a flag in addition, sockets that are waiting for a connection (i.e., listening) are displayed as well. This display will give you a list of all servers that are currently running on your system

Invoking **netstat -ta** on vlager produces this output:

```
$ netstat -ta
Active Internet Connections
Proto Recv-Q Send-Q Local Address   Foreign Address       (State)
tcp    0      0 *:domain            *:*                   LISTEN
tcp    0      0 *:time              *:*                   LISTEN
tcp    0      0 *:smtp              *:*                   LISTEN
tcp    0      0 vlager:smtp         vstout:1040           ESTABLISHED
tcp    0      0 *:telnet            *:*                   LISTEN
tcp    0      0 localhost:1046      vbardolino:telnet     ESTABLISHED
tcp    0      0 *:chargen           *:*                   LISTEN
tcp    0      0 *:daytime           *:*                   LISTEN
tcp    0      0 *:discard           *:*                   LISTEN
tcp    0      0 *:echo              *:*                   LISTEN
tcp    0      0 *:shell             *:*                   LISTEN
tcp    0      0 *:login             *:*                   LISTEN
```

This output shows most servers simply waiting for an incoming connection. However, the fourth line shows an incoming SMTP connection from vstout, and the sixth line tells you there is an outgoing **telnet** connection to vbardolino.

2.13.5. Ipconfig (Internet Protocol Configuration)

ipconfig (*internet protocol* ***configuration***) in Microsoft Windows is a console application that displays all current TCP/IP network configuration values and refreshes Dynamic Host Configuration Protocol DHCP and Domain Name System DNS settings

Syntax

```
ipconfig [/all] [/renew [Adapter]] [/release [Adapter]] [/flushdns] [/displaydns] [/registerdns] [/showclassid Adapter] [/setclassid Adapter [ClassID]]
```

Parameters

/all : Displays the full TCP/IP configuration for all adapters. Without this parameter, **ipconfig** displays only the IP address, subnet mask, and default gateway values for each adapter. Adapters can represent physical interfaces, such as installed network adapters, or logical interfaces, such as dial-up connections.

/renew [*Adapter*] : Renews DHCP configuration for all adapters (if an adapter is not specified) or for a specific adapter if the *Adapter* parameter is included. This parameter is available only on computers with adapters that are configured to obtain an IP address automatically. To specify an adapter name, type the adapter name that appears when you use **ipconfig** without parameters.

/release [*Adapter*] : Sends a DHCPRELEASE message to the DHCP server to release the current DHCP configuration and discard the IP address configuration for either all adapters (if an adapter is not specified) or for a specific adapter if the *Adapter* parameter is included. This parameter disables TCP/IP for adapters configured to obtain an IP address automatically. To specify an adapter name, type the adapter name that appears when you use **ipconfig** without parameters.

/flushdns : Flushes and resets the contents of the DNS client resolver cache. During DNS troubleshooting, you can use this procedure to discard negative cache entries from the cache, as well as any other entries that have been added dynamically.

/displaydns : Displays the contents of the DNS client resolver cache, which includes both entries preloaded from the local Hosts file and any recently obtained resource records for name queries resolved by the computer. The DNS Client service uses this information to resolve frequently queried names quickly, before querying its configured DNS servers.

/registerdns : Initiates manual dynamic registration for the DNS names and IP addresses that are configured at a computer. You can use this parameter to troubleshoot a failed DNS name registration or resolve a dynamic update problem between a client and the DNS server without rebooting the client computer. The DNS settings in the advanced properties of the TCP/IP protocol determine which names are registered in DNS.

/showclassid *Adapter* : Displays the DHCP class ID for a specified adapter. To see the DHCP class ID for all adapters, use the asterisk (*) wildcard character in place of *Adapter*. This parameter is available only on computers with adapters that are configured to obtain an IP address automatically.

/setclassid *Adapter* [*ClassID*] : Configures the DHCP class ID for a specified adapter. To set the DHCP class ID for all adapters, use the asterisk (*) wildcard character in place of *Adapter*. This parameter is available only on computers with adapters that are configured to obtain an IP address automatically. If a DHCP class ID is not specified, the current class ID is removed.

/?: Displays help at the command prompt

Key Points

1. The **ipconfig** command is the command-line equivalent to the **winipcfg** command, which is available in Windows Millennium Edition, Windows 98, and Windows 95. Although Windows XP does not include a graphical equivalent to the **winipcfg** command, you can use Network Connections to view and renew an IP address. To do this, open Network Connections, right-click a network connection, click **Status**, and then click the **Support** tab.
2. This command is most useful on computers that are configured to obtain an IP address automatically. This enables users to determine which TCP/IP configuration values have been configured by DHCP, Automatic Private IP Addressing (APIPA), or an alternate configuration.
3. If the *Adapter* name contains any spaces, use quotation marks around the adapter name (that is, "*Adapter Name*").
4. For adapter names, **ipconfig** supports the use of the asterisk (*) wildcard character to specify either adapters with names that begin with a specified string or adapters with names that contain a specified string. For example, **Local*** matches all adapters that start with the string Local and ***Con*** matches all adapters that contain the string Con.
5. This command is available only if the **Internet Protocol (TCP/IP)** protocol is installed as a component in the properties of a network adapter in Network Connections .

Example 1. Write command To display the basic TCP/IP configuration for all adapters, type:

Solution. ipconfig

Example 2. Write command To display the full TCP/IP configuration for all adapters, type:

Solution. ipconfig /all

Example 3. Write command To renew a DHCP-assigned IP address configuration for only the **Local Area Connection** adapter, type:

Solution. ipconfig /renew "Local Area Connection"

Example 4. Write command to flush the DNS resolver cache when troubleshooting DNS name resolution problems, type.

Solution. ipconfig /flushdns

Example 5. Write command To display the DHCP class ID for all adapters with names that start with *Local*, type:

Solution. ipconfig /showclassid Local*

Example 6. Write command To set the DHCP class ID for the **Local Area Connection** adapter to *TEST*, type:

Solution. ipconfig /setclassid "Local Area Connection" TEST

Formatting legend

Format	Meaning
Italic	Information that the user must supply
Bold	Elements that the user must type exactly as shown
Ellipsis (...)	Parameter that can be repeated several times in a command line
Between brackets ([])	Optional items
Between braces ({}); choices separated by pipe (l). Example: {evenlodd}	Set of choices from which the user must choose only one
Courier font	Code or program output

HIGHLIGHTS

1. OSI model has seven layers :
   - ➢ Application — Top most
   - ➢ Presentation
   - ➢ Session
   - ➢ Transport
   - ➢ Network
   - ➢ Data link
   - ➢ Physical — Lower most
2. TCP/IP protocol architecture has five layers :
   - ➢ Application
   - ➢ Transport
   - ➢ Network
   - ➢ Data link
   - ➢ Physical
3. Address at physical layer are physical address.
4. Address at network layer are logical or IP address.

5. Address at transport layer are port address.
6. Various version of IP are available :
   - IPV4
   - IPV5
   - IPV6
7. IP address is 32 bits address that uniquely defines the connection of a host or router.
8. IP address may be classfull or classless.
9. Three notations of IP address are possible.
   - Binary notation
   - Decimal notation
   - Hexadecimal notation
10. In classfull addressing is divided into five classes
    - Class A
    - Class B
    - Class C
    - Class D
    - Class E
11. Class A use 50% address space, class B used up 25% class, class C used up 12.5% and class D and E used up 6.25% of address space.
12. IP address is divided into :
    - Network ID (network address)
    - Hostid (host address)
13. Disadvantage of classfull address is that it wastes a lot of IP addresses.
14. In classless addressing the entire range of IP addresses is divided into blocks of variable sizes.
15. Format of classess IP address is x.y.x.t/h
16. Subnetting is the process of designating some high-order bits from the host port and grouping them with the network mask to form the subnet mask.
17. Subnet mask extracts network address from IP address.
18. Subnet mask is of two types :
    - Default mask
    - Subnet mask
19. Supernetting is a way to aggregate multiple internet addresses of the same class.
20. Supernetting is also called classless inter-domain routing (CIDR).
21. ARP protocol converts logical address to physical address.
22. RARP converts logical address to physical address.
23. ICMP protocol is a helper protocol that supports IP with facility for error reporting simple queries.

QUESTIONS WITH ANSWERS

Q. 1. What is difference between TCP and UDP

Ans. **TCP :** It is connection oriented protocol. TCP is sophisticated reliable byte stream Protocol.

UDP : It is a connectionless protocol. UDP is simple unreliable datagram protocol.

Q. 2. What is the use of ARP and RARP?

Ans. **ARP (Address Resolution Protocol) :** It is used to find the physical address if the IP address is known.

RARP (Reverse Address Resolution Protocol) : It is used to find the IP address if the physical address is known.

Q. 3. What is MAC address?

Ans. The address for a device as it is identified at the Media Access Control (MAC) layer in the network architecture. MAC address is usually stored in ROM on the network adapter card and is unique.

Q. 4. What is the difference between physical address and logical address?

Ans. A Physical address is a 48-bit flat address burned into the ROM of the NIC card which is a Layer 1 device of the OSI model. This iis divided into 24-bit vendor code and 24-bit serial address. This is unique for each system and cannot be changed.

A Logical address is a 32-bit address assigned to each system in a network. This works in Layer-3 of OSI Model. This would be generally the IP address.

Q. 5. What are the Advantages and Disadvantages of DHCP?

Ans. **Advantages :** All the IP configuration information gets automatically configured for your client machine by the DHCP server.

If you move your client machine to a different subnet, the client will send out its discover message at boot time and work as usual. However, when you first boot up there you will not be able to get back the IP address you had at your previous location regardless of how little time has passed.

Disadvantage : Your machine name does not change when you get a new IP address. The DNS (Domain Name System) name is associated with your IP address and therefore does change. This only presents a problem if other clients try to access your machine by its DNS name.

Q. 6. Describe a 3-way TCP/IP Handshake.

Ans. To establish a connection, TCP uses a 3-way handshake. Before a client attempts to connect with a server, the server must first bind to a port to open it up for connections: this is called a passive open. Once the passive open is established then a client may initiate an active open. To establish a connection, the 3-way (or 3-step) handshake occurs:

The active open is performed by sending a SYN to the server.

In response, the server replies with a SYN-ACK.

Finally the client sends an ACK back to the server.

At this point, both the client and server have received an acknowledgement of the connection.

Q. 7. What is the difference between TFTP and FTP application layer protocols?

Ans. The Trivial File Transfer Protocol (TFTP) allows a local host to obtain files from a remote host but does not provide reliability or security. It uses the fundamental packet delivery services offered by UDP.

The File Transfer Protocol (FTP) is the standard mechanism provided by TCP/IP for copying a file from one host to another. It uses the services offered by TCP and so is reliable and secure. It establishes two connections (virtual circuits) between the hosts, one for data transfer and another for control information.

Q. 8. What is mean by DNS (Domain Name System) and what is the purpose of IT ?

Ans. DNS allows you to connect to another networked computer or remote service by using its user-friendly domain name rather than its numerical IP address. Conversely, Reverse DNS (rDNS) translates an IP address into a domain name.A DNS server provides the name resolution service required by Active Directory.DNS is used by many applications to resolve Internet hostnames into IP addresses.

Q. 9. What is RIP (Routing Information Protocol)?

Ans. It is a simple protocol used to exchange information between the routers.

Q. 10. What is supernetting? Why it is used ?

Ans. Supernetting refers to increase host and reduce the subnet's. It mean reduce the network bits and increase the host bits. We do use supernetting because when the PCs in the network increase where we need the more valid IP's in the same network.

REVIEW QUESTIONS

1. Explain the architecture of TCP/IP protocol suite.
2. Explain the function of each layer in OSI.
3. What is difference between ARP and RARP.
4. What is IP address.
5. What is the difference between classless addressing and classful addressing.
6. What is subnetting ?
7. What is supernetting ?
8. What is subnet mask ?
9. Discuss default subnet mask.
10. What is the difference between port address, physical address and IP address.

❑❑❑

IP Datagram Format

INSIDE THIS CHAPTER

Chapter **3**

3.1. INTRODUCTION

The term 'datagram' or 'packet' is used to describe a chunk of IP data. Each IP datagram contains a specific set of fields in a specific order so that the reader knows how to decode and read the stream of data received.

> **DO YOU KNOW**
>
> **IP datagram is aiso kown as IPV4 datagram.**

Format of IP datagram is as follows :

0	4	8	12	16	20	24	28
Version	IHL	Service Type		Total Length			
Identification				Flags	Fragment Offset		
Time to Live		Protocol		Header Checksum			
Source address							
Destination address							
Options + Padding (0 or more 32-bit words)							
Data							
More data...							

Fig. 3.1. IPV4 Datagram Format.

The various fields are described below :

1. Version (4 Bits)

The version field is set to the value '4' in decimal or '0100' in binary. The value indicates the version of IP (4 or 6, there is no version 5).

2. IHL (4 Bits)

The Internet Header Length (IHL) describes how big the header is in 32-bit words. For instance, the minimum value is 5, as that is the minimum size of an IP header that contains all the correct fields is 160 bits, or 20 bytes. This allows the receiver to know exactly where the payload data begins.

3. TOS (8 Bits)

Type of service allows the intermediate receiving stations (the routers) to have some notion of the quality of service desired. This allows the network to make adaptations for delay, throughput, or reliability.

4. Total Length (16 Bits)

This informs the receiver of the datagram where the end of the data in this datagram is. This is the length of the entire datagram in octets, including the header. This is why an IP datagram can be up to 65,535 bytes long, as that is the maximum value of this 16-bit field.

5. Identification (16 Bits)

Sometimes, a device in the the middle of the network path cannot handle the datagram at the size it was originally transmitted, and must break it into fragments. If an intermediate system needs to break up the datagram, it uses this field to aid in identifying the fragments.

6. Flags (3 Bits)

The flags field contains single-bit flags that indicate whether the datagram is a fragment, whether it is permitted to be fragmented, and whether the datagram is the last fragment, or there are more fragments. The first bit in this field is always zero.

7. Fragment Offset (13 Bits)

When a datagram is fragmented, it is necessary to reassemble the fragments in the correct order. The fragment offset numbers the fragments in such a way that they can be reassembled correctly.

8. Time to Live (8 Bits)

This field determines how long a datagram will exist. At each hop along a network path, the datagram is opened and it's time to live field is decremented by one (or more than one in some cases). When the time to live field reaches zero, the datagram is said to have 'expired' and is discarded. This prevents congestion on the network that is created when a datagram cannot be forwarded to it's destination. Most applications set the time to live field to 30 or 32 by default.

9. Protocol (8 Bits)

This indicates what type of protocol is encapsulated within the IP datagram. Some of the common values seen in this field include.

Protocol	Number (Decimal)
ICMP	1
IGMP	2
TCP	6
UDP	17

10. Header Checksum (16 Bits)

According to RFC 791, the header checksum formula is: *"the 16-bit ones compliment of the ones compliment sum of all 16-bit words in the header."*

The checksum allows IP to detect datagrams with corrupted headers and discard them. Since the time to live field changes at each hop, the checksum must be re-calculated at each hop. In some cases, this is replaced with a cyclic redundancy check algorithm.

11. Source Address (32 Bits)

This is the IP address of the sender of the IP datagram.

12. Destination Address (32 bits)

This is the IP address of the intended receiver(s) of the datagram. If the host portion of this address is set to all 1's, the datagram is an 'all hosts' broadcast.

13. Options and Padding (Variable)

Various options can be included in the header by a particular vendor's implementation of IP. If options are included, the header must be padded with zeroes to fill in any unused octets so that the header is a multiple of 32 bits, and matches the count of bytes in the Internet Header Length (IHL) field.

3.2. UDP DATAGRAM FORMAT [MDU Dec. 2008; May 2006]

It provides a connectionless, best-effort packet delivery service. As UDP provides no guarantees to the upper layer protocol for message delivery hence it is also referred to as

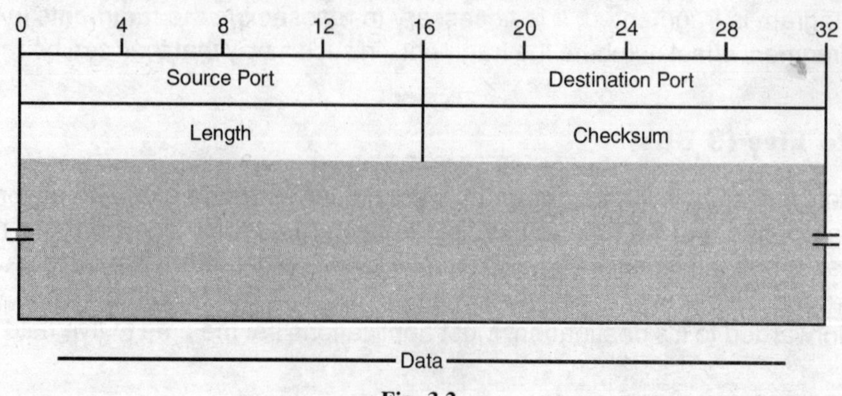

Fig. 3.2.

Unreliable Datagram Protocol. Any reliability issues for the information must be implemented and handled in upper layers.

The UDP header consists of only 4 fields.

Source Port

This field identifies the sending port when meaningful and should be assumed to be the port to reply to if needed. This field is optional, when not used it should be zero.

Destination Port

This field identifies the destination port and is mandatory.

Length

A 16-bit field that specifies the length in bytes of the entire datagram: header and data in total. The minimum length is 8 bytes since that is the length of the header.

Checksum

The 16-bit checksum field is used for error-checking of the header and data. This field is optional.

Checksum is the 16-bit one's complement of the one's complement sum of a pseudo header of information from the IP header, the UDP header, and the data, padded with zero octets at the end (if necessary) to make a multiple of two octets.

Data

This field contains the actual data.

3.3. TCP HEADER FORMAT [MDU Dec. 2008; May 2006]

The Transmission Control Protocol, provides a reliable, connection-oriented service. In other words, TCP provides a point-to-point connection.

> **DO YOU KNOW**
>
> **A packet in TCP is called a segment.**

Point-to-point connections have two characteristics :

- They have only one path to the destination. A packet entering one end of the connection cannot become lost, because the only place to go is the other end.
- Packets arrive in the same order in which they are sent.
- The internet layer TCP uses a connectionless, best-effort packet delivery service. The analogy of this is the Postal Service. If a stack of letters is given to the mail carrier for delivery, there is no guarantee that the letters will arrive stacked in the same order, that they will all arrive on the same day, or indeed that they will arrive at all. The Postal Service merely commits to making its best effort to deliver the letters.
- Likewise, the internet layer does not guarantee that all packets will take the same route, and therefore there is no guarantee that they will arrive in the same sequence and time intervals as they were sent, or that they will arrive at all.
- On the other hand, a telephone call is connection-oriented service. Data must arrive sequentially and reliably, or it is useless. Like a telephone call, TCP must first establish

a connection, then transfer data, and then perform a disconnect when the data transfer is complete.

➢ TCP uses three fundamental mechanisms to accomplish a connection-oriented service on top of a connectionless service:

➢ Packets are labeled with sequence numbers so that the receiving TCP service can put out-of-sequence packets into the correct sequence before delivering them to the destination application.

➢ TCP uses a system of acknowledgments, checksums, and timers to provide reliability. A receiver might notify a sender when it recognizes that a packet in a sequence has failed to arrive or has errors, or a sender might assume that a packet has not arrived if the receiver does not send an acknowledgment within a certain amount of time after transmission. In both cases, the sender will resend the packet in question.

➢ TCP uses a mechanism called windowing to regulate the flow of packets; windowing decreases the chances of packets being dropped because of full buffers in the receiver.

➢ TCP attaches a header to the application layer data; the header contains fields for the sequence numbers and other information necessary for these mechanisms fields for addresses called port numbers, which identify the source and destination applications of the data. The application data with its attached TCP header is then encapsulated within an IP packet for delivery.

TCP data is encapsulated in an IP datagram. The figure shows the format of the TCP header. Its normal size is 20 bytes unless options are present. Each of the fields is discussed below:

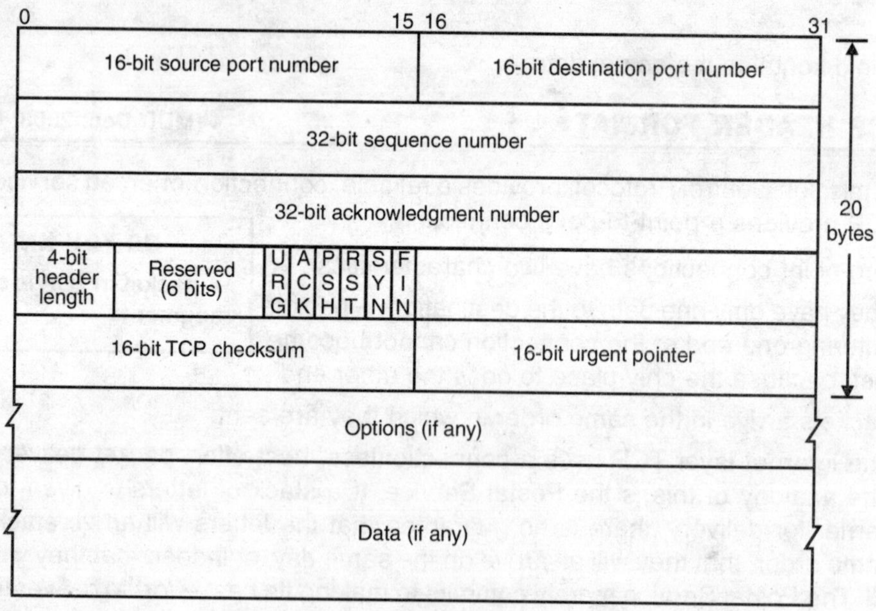

Fig. 3.3. TCP Header Format

The various fields in TCP are described below :

1. The **SrcPort** and **DstPort** fields identify the source and destination ports,respectively. These two fields plus the source and destination IP addresses, combine to uniquely identify each TCP connection.

2. The **sequence number** identifies the byte in the stream of data from the sending TCP to the receiving TCP that the first byte of data in this segment represents.
3. The **Acknowledgement number** field contains the next sequence number that the sender of the acknowledgement expects to receive. This is therefore the sequence number plus 1 of the last successfully received byte of data. This field is valid only if the ACK flag is on. Once a connection is established the Ack flag is always on.
4. The **Acknowledgement**, **SequenceNum**, and **AdvertisedWindow** fields are all involved in TCP's sliding window algorithm. The Acknowledgement and AdvertisedW indow fields carry information about the flow of dat going in the other direction. In TCP's sliding window algorithm the reciever advertises a window size to the sender. This is done using the AdvertisedWindow field. The sender is then limited to having no more than a value of AdvertisedWindow bytes of un acknowledged data at any given time. The receiver sets a suitable value for the AdvertisedWindow based on the amount of memory allocated to the connection for the purpose of buffering data.
5. The **header length** gives the length of the header in 32-bit words. This is required because the length of the options field is variable.
6. The 6-bit **Flags field** is used to relay control information between TCP peers. The possible flags include SYN, FIN, RESET, PUSH, URG, and ACK.
   - ➢ The SYN and Fin flags are used when establishing and terminating a TCP connection, respectively.
   - ➢ The ACK flag is set any time the Acknowledgement field is valid, implying that the receiver should pay attention to it.
   - ➢ The URG flag signifies that this segment contains urgent data. When this flag is set, the UrgPtr field indicates where the non-urgent data contained in this segment begins.
   - ➢ The PUSH flag signifies that the sender invoked the push operation, which indicates to the receiving side of TCP that it should notify the receiving process of this fact.
   - ➢ Finally, the RESET flag signifies that the receiver has become confused and so wants to abort the connection.
7. The **Checksum** covers the TCP segment: the TCP header and the TCP data. This is a mandatory field that must be calculated by the sender, and then verified by the receiver.
8. The **Option field** is the maximum segment size option, called the MSS. Each end of the connection normally specifies this option on the first segment exchanged. It specifies the maximum sized segment the sender wants to recieve.
9. The **data** portion of the TCP segment is optional.

DO YOU KNOW

The TCP header can have up to 40 bytes of optional information. Options convey additional information to the destination or align other options.

TIPS

The inclusion of the checksum in TCP is mandatory.

3.3.1. Difference Between TCP and UDP Datagram

Transmission Control Protocol (TCP)

(i) Transmission Control Protocol (TCP) is a connection oriented protocol, which means the devices should open a connection before transmitting data and should close the connection gracefully after transmitting the data.
(ii) Transmission Control Protocol (TCP) assure reliable delivery of data to the destination.
(iii) Transmission Control Protocol (TCP) protocol provides extensive error checking mechanisms such as flow control and acknowledgment of data.
(iv) Sequencing of data is a feature of Transmission Control Protocol (TCP).
(v) Delivery of data is guaranteed if you are using Transmission Control Protocol (TCP).
(vi) Transmission Control Protocol (TCP) is comparatively slow because of these extensive error checking mechanisms
(vii) Multiplexing and Demultiplexing is possible in Transmission Control Protocol (TCP) using TCP port numbers.
(viii) Retransmission of lost packets is possible in Transmission Control Protocol (TCP).

User Datagram Protocol (UDP)

(i) User Datagram Protocol (UDP) is Datagram oriented protocol with no overhead for opening, maintaining, and closing a connection.
(ii) User Datagram Protocol (UDP) is efficient for broadcast/multicast transmission.
(iii) User Datagram protocol (UDP) has only the basic error checking mechanism using checksums.
(iv) There is no sequencing of data in User Datagram protocol (UDP).
(v) The delivery of data cannot be guaranteed in User Datagram protocol (UDP).
(vi) User Datagram protocol (UDP) is faster, simpler and more efficient than TCP. However, User Datagram protocol (UDP) it is less robust then TCP.
(vii) Multiplexing and Demultiplexing is possible in User Datagram Protocol (UDP) using UDP port numbers.
(viii) There is no retransmission of lost packets in User Datagram Protocol (UDP).

Table 3.1. Difference between TCP and UDP Datagram

S. No.	TCP	UDP
1.	*Reliability* : TCP is connection-oriented protocol. When a file or message send it will get delivered [unless connections fails. If connection lost, the server will request the lost part. There is no corruption while transferring a message.	*Reliability:* UDP is connectionless protocol. When you a send a data or message, you don't know if it'll get there, it could get lost on the way. There may be corruption while transferring a message.
2.	*Ordered:* If you send two messages along a connection, one after the other, you know the first message will get there first. You don't have to worry about data arriving in the wrong order.	*Ordered:* If you send two messages lout, you don't know what order (they'll arrive in i.e. **no ordered.**

S. No.	TCP	UDP
3.	*Heavyweight:* When the low level parts of the TCP "stream" arrive in the wrong order, resend requests have to be sent, and all the out of sequence parts have to be put back together, so requires a bit of work to piece together.	*Lightweight :* No ordering of messages, no tracking connections, etc. It's just fire and forget! This means it's a lot quicker, and the network card/OS have to do very little work to translate the data back from the packets.
4.	*Streaming:* Data is read as a "stream," with nothing distinguishing where one packet ends and another begins. There may be multiple (packets per read call.	*Datagrams:* Packets are sent individually and are guaranteed to be whole if they arrive. One packet per lone read call.
5.	*Examples:* World Wide Web (Apache TCP port 80), e-mail (SMTP TCP port 25 Postfix MTA), File Transfer Protocol (FTP port 21) (and Secure Shell (OpenSSH port 22) etc.	*Examples:* Domain Name System (DNS UDP port 53), streaming media applications such as IPTV or movies, Voice over IP (VoIP), Trivial File Transfer Protocol (TFTP) and online multiplayer games etc.

3.4. IPV6 HEADER FORMAT

[MDU Dec. 2007]

IPv6 consists of following fields

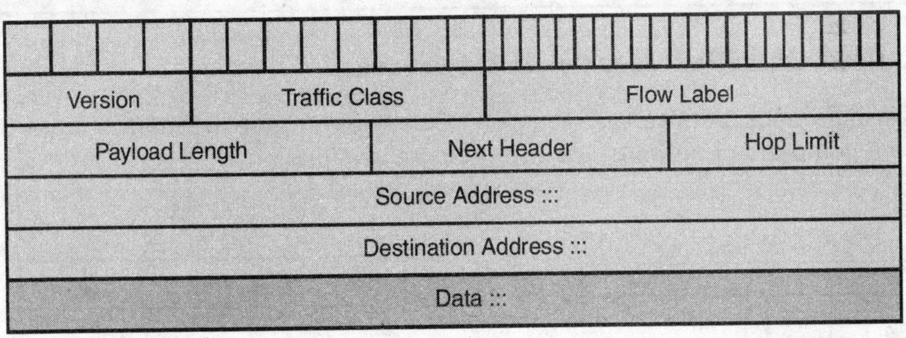

Fig. 3.4.

The fields are described as

Version. 4 bits.

IPv6 version number.

Traffic Class. 8 bits.

Internet traffic priority delivery value.

Flow Label. 20 bits.

Used for specifying special router handling from source to destination(s) for a sequence of packets.

Payload Length. 16 bits unsigned.

Specifies the length of the data in the packet. When cleared to zero, the option is a hop-by-hop Jumbo payload.

Next Header. 8 bits.

Specifies the next encapsulated protocol. The values are compatible with those specified for the IPv4 protocol field.

Hop Limit. 8 bits unsigned.

For each router that forwards the packet, the hop limit is decremented by 1. When the hop limit field reaches zero, the packet is discarded. This replaces the TTL field in the IPv4 header that was originally intended to be used as a time based hop limit.

Source address. 16 bytes.

The IPv6 address of the sending node.

Destination address. 16 bytes.

The IPv6 address of the destination node.

TIPS

An IPv6 packet is a block of data that contains a header and a payload. The header is the information necessary to deliver the packet to a destination address; the payload is the data that you want to deliver. IPv6 packets can use a standard or an extended format.

3.4.1. IPv4 and IPv6 Header Differences

The main difference between IPv4 and IPv6 resides in their headers. Figure 3.5 below provides a comparison between the two protocol versions.

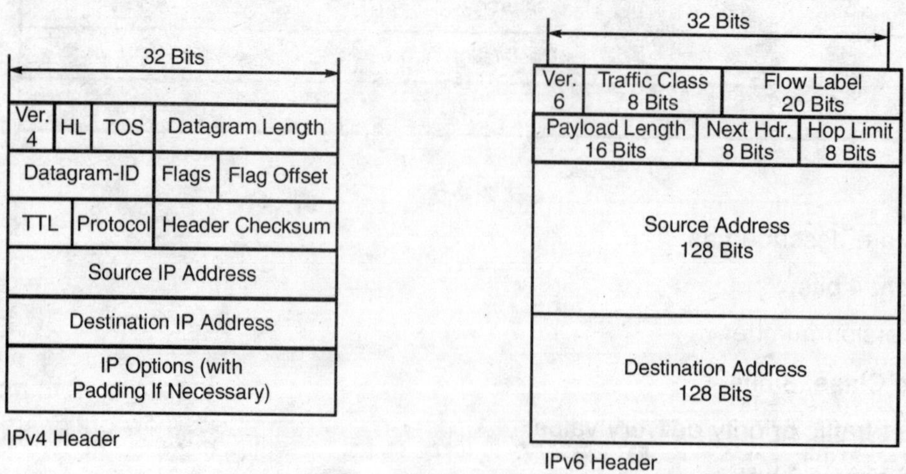

Fig. 3.5. IPv4 and IPv6 Header Comparison.

Standard IPv6 Headers

IPv6 packet headers contain many of the fields found in IPv4 packet headers; some of these fields differ from IPv4.

The 40-byte IPv6 header consists of the following eight fields:

- **Version -** Indicates the version of the Internet Protocol.
- **Traffic class -** Previously the type-of-service (ToS) field in IPv4, the traffic class field defines the class-of-service (CoS) priority of the packet. However, the semantics for this field (for example, diffserv code points) are identical to IPv4.
- **Flow label -** The flow label identifies all packets belonging to a specific flow (that is, packet flows requiring a specific class of service [CoS]); routers can identify these packets and handle them in a similar fashion.
- **Payload length -** Previously the total length field in IPv4, the payload length field specifies the length of the IPv6 payload.
- **Next header -** Previously the protocol field in IPv4, the Next Header field indicates the next extension header to examine.
- **Hop limit -** Previously the time-to-live (TTL) field in IPv4, the hop limit indicates the maximum number of hops allowed.
- **Source address -** Identifies the address of the source node sending the packet.
- **Destination address -** Identifies the final destination node address for the packet.

Extension Headers

In IPv6, extension headers are used to encode optional Internet-layer information. Extension headers are placed between the IPv6 header and the upper-layer header in a packet.

IPv6 allows you to chain extension headers together by using the next header field. The next header field, located in the IPv6 header, indicates to the router which extension header to expect next. If there are no more extension headers, the next header field indicates the upper-layer header (TCP header, UDP header, ICMPv6 header, an encapsulated IP packet, or other items).

Internet Protocol version 6 (IPv6) is designed to eventually supersede IP version 4 (IPv4). The intent of this design change is not to take a radical step away from IPv4, but to enhance IP addressing and maintain other IPv4 functions that work well.

The differences between IPv4 and IPv6 include the following :

- **Expanded addressing capabilities :** IPv6 increases the size of the IP address from 32 bits to 128 bits. This increased size allows for a larger address space and a much larger number of addressable nodes.
- **Simplified header format :** Reducing some common processing costs associated with packet handling and streamlining the bandwidth cost of the larger IPv6 header, some IPv4-specific header fields either no longer exist or are now optional in the IPv6 header.
- **Traffic flow labelling capabilities :** The ability to label packets for specific traffic flows exists in the IPv6 packet. These labels allow for nondefault quality of service (QOS) or the possibility of "real-time" services.

- **Authentication capabilities :** Authentication provides the ability to use extensions for some authentication and data integrity applications.

  IPv6 continues to provide the basic packet delivery service for all TCP/IP networks. As a coiinectionless protocol, IPv6 does not exchange control information to establish an end-tc-end connection before transmitting data. Instead, just like its IPv4 predecessor, IPv6 continues to' rely on protocols in other layers to establish the connection if connection-oriented services are required and to provide error detection and error recovery.

3.5. FINGER PROTOCOL [MDU Dec. 2007]

In computer networking, the Name/Finger protocol and the Finger user information protocol are simple network protocols for the exchange of human-oriented status and user information.

If we execute the finger command specifying a user at a particular host, we will obtain specific information about that user. The FINGER Protocol must be present at the remote host and at the requesting host. FINGER uses Transmission Control Protocol (Transmission Control Protocol) as its underlying protocol.

To use the Finger protocol, a client would send a one-line query to the Finger server at a site. The server would send back the answer and close the connection.

DO YOU KNOW

The finger protocol is a simple network protocol based on RFC1288.

TIPS

The Finger was used in the 1980s and 1990s to identify users on a host

The server side of the protocol is implemented by a program "fingerd", while the client side is implemented by the "name" and "finger" programs.

The program is supposed to return a friendly, human-oriented status report on either the system at the moment or a particular person in depth. There is no required format, and the protocol consists mostly of specifying a single command line. It is most often implemented on Unix or Unix-like systems.

The program would supply information such as whether a user is currently logged-on, e-mail address, full name etc. As well as standard user information, finger displays the contents of ".plan" file in the user's home directory. Often this file (maintained by the user) contained either useful information about the user's current activities, or alternatively all manner of humor.

Supplying such detailed information was considered acceptable and convenient in the early days of the Internet, but later was considered questionable for privacy and security reasons. Finger information has been frequently used by crackers as a way to initiate a social engineering attack on a company's computer security system. By using a finger client to get a list of a company's employee names, email addresses, phone numbers, and so on, a cracker can telephone or email someone at a company requesting information while posing as another employee. Information garnered from finger can be used both to find an employee to contact directly, and to more effectively pose as an existing employee.The finger protocol has less security aspect.

3.5.1. Uses of Finger Protocol

The Finger protocol could be used to reveal information about users like their login names, phone numbers and whether they were logged on at a certain time. The protocol could also be used to send email to a user

3.6. IPCONFIG COMMAND [MDU Dec. 2007]

ipconfig is a commmand line utility available on all versions of Microsoft Windows starting with Windows NT. ipconfig is designed to be run from the Windows command prompt. This utility allows you to get the IP address information of a Windows computer. It also allows some control over active TCP/IP connections. ipconfig is an alternative to the older 'winipcfg' utility.

If we type ipconfig in command prompt with default options,The output of the default command contains the IP address, network mask and gateway for all physical and virtual network adapters.

'ipconfig' supports several command line options as described below. The command "ipconfig /?" displays the set of available options.

1. ipconfig /all

This option displays the same IP addressing information for each adapter as the default option. Additionally, it displays DNS and WINS settings for each adapter.

2. ipconfig /release

This option terminates any active TCP/IP connections on all network adapters and releases those IP addresses for use by other applications. 'ipconfig /release" can be used with specific Windows connection names. In this case, the command will affect only the specified connections and not all. The command accepts either full connection names or wildcard names. Examples:

```
ipconfig /release "Local Area Connection 1"
ipconfig /release *Local*
```

3. ipconfig /renew

This option re-establishes TCP/IP connections on all network adapters. As with the release option, ipconfig /renew takes an optional connection name specifier.

Both /renew and /release options only work on clients configured for dynamic (DHCP) addressing.

4. ipconfig /showclassid, ipconfig /setclassid

These options manage DHCP class identifiers. DHCP classes can be defined by administrators on a DHCP server to apply different network settings to different types of clients. This is an advanced feature of DHCP typically used in business networks, not home networks.

5. ipconfig /displaydns, ipconfig /flushdns

These options access a local DNS cache that Windows maintains. The /displaydns option prints the contents of the cache, and the /flushdns option erases the contents.

This DNS cache contains a list of remote server names and the IP addresses (if any) they correspond to. Entries in this cache come from DNS lookups that happen when attempting to visit Web sites, named FTP servers, and other remote hosts. Windows uses this cache to improve the performance of Internet Explorer and other Web-based applications.

TIPS

IPconfig is used primarily to troubleshoot internet connections.

In home networking, these DNS options are sometimes useful for advanced troubleshooting. If the information in your DNS cache becomes corrupted or outdated, you could face difficulty accessing certain sites on the Internet. Consider these two scenarios:

- ➢ The IP address of a Web site, email server or other server changes (rare occurence). The name and address of this site normally stay in your cache for 24 hours after your last visit. You may need to clear your cache to access the server sooner.
- ➢ A Web site or other server was offline when you last visited it (hopefully a rare occurence) but since has come back online. The cache will normally keep a record that the server is offline for 5 minutes afer your last visit. You may need to clear your cache to access the server sooner.

6. ipconfig /registerdns

Similar to the above options, this option updates DNS settings on the Windows computer. Instead of merely accessing the local DNS cache, however, this option initiates communication with both the DNS server (and the DHCP server) to re-register with them.

This option is useful in troubleshooting problems involving connection with the Internet service provider, such as failure to obtain a dynamic IP address or failure to connect to the ISP DNS server.

DO YOU KNOW

The ipconfig.exe command is an external command and is available in the below Microsoft operating systems.

Like the /release and /renew options, /registerdns optionally takes the name(s) of specific adapters to update. If no name parameter is specified, /registerdns updates all adapters.

3.7. PING COMMAND [MDU Dec. 2007]

Ping is a basic Internet program that allows a user to verify that a particular IP address exists and can accept requests.

OR

Ping is the name of a standard software utility (tool) used to test network connections. It can be used to determine if a remote device (such as Web or game server) can be reached across the network and, if so, the connection's latency.

Ping tools are part of Windows, Mac OS X and Linux as well as some routers and game consoles.The term latency refers to any of several kinds of delays typically incurred in processing of network data. A low latency network connection is one that generally experiences small delay times, while a high latency connection generally suffers from long delays.

TIPS

PING is used primarily to troubleshoot Internet connections

DO YOU KNOW

Full form of ping is Packet Internet Groper.

Ping is used diagnostically to ensure that a host computer the user is trying to reach is actually operating. Ping works by sending an Internet Control Message Protocol (ICMP) Echo Request to a specified interface on the network and waiting for a reply. Ping can be used for troubleshooting to test connectivity and determine response time.

The output of ping varies depending on the tool. Standard results includes

- IP address of the responding computer.
- Length of time (in milliseconds) between sending the request and receiving the response.
- An indication of how many network hops between the requesting and responding computers.
- Error messages if the target computer did not respond.

3.8. TRACERT COMMAND (TRACEROUTE) [MDU May 2008]

If you are having connectivity problems, you can use the tracert command to check the path to the destination IP address that you want to reach and record the results. The tracert command displays the series of IP routers that are used in delivering packets from your computer to the destination and how long it took on each hop. If the packets are unable to be delivered to the destination, the tracert command displays the last router that successfully forwarded your packets.

TIPS

The traceroute command traces the network path of Internet routers that packets take as they are forwarded from your computer to a destination address.

TIPS

Traceroutes can be useful to diagnose slow network connections.

Type **tracert** at a command prompt.

The most common use of **tracert** is as follows:

```
tracert IP address [-d]
```

This returns a list of the routers that are crossed to get to *IP address*. By using the **-d** option, the router path is displayed faster because **tracert** does not try to resolve the names of the routers in the path.

DO YOU KNOW

On Unix and Linux computers, the "traceroute" application is available in the shell, while on Windows computers, the "tracert" program can be accessed from DOS.

3.9. ROUTE COMMAND

Route is a tremendously useful command for manipulating and displaying IP routing tables. Various task can be performed with route command

- ➢ Display routes,
- ➢ Add routes (most importantly, the default route),
- ➢ Remove routes,
- ➢ Examine the routing cache

3.9.1. Displaying the Routing Table With Route

route helps in viewing the routing table. It is by using the route command that you can determine where a packet goes when it leaves your machine.

HIGHLIGHTS

1. The term datagram or packet is used to describe a chunck of IP data.
2. TCP is connection-oriented protocol.
3. UDP is connectionless protocol.
4. The difference between IPv4 and IPv6 is IPv6 increases the size of IP address from 32 bits to 128 bits.
5. IPv6 has simpler header format in comparison to IPv4.
6. Finger protocol is used to identify users on a host.
7. IPconfig allow to get the IP address information of a window computer.
8. Ping is used primarily to troubleshoot internet connections.
9. Trace route command can be useful to diagnose slow network connection.
10. Route is useful command for manipulating and displaying IP routing tables.

QUESTIONS WITH ANSWERS

Q. 1. What Is a Computer Ping Test?

Ans. *Ping test* determines whether your computer can communicate with another computer over the network. Then, if network communication is established, ping tests also determine the connection latency (delay) between the two computers.

We can use a ping test to troubleshoot connectivity problems with your home network. Ping tests are also commonly used to measure the delay with online gaming services or other Internet servers.

Microsoft Windows contains a built-in *ping utility* for running ping tests. Alternative utilities are also freely available as Web downloads. To execute a ping test, you simply identify the Web site or other remote server / computer by its name or IP address. The result of a ping test includes confirmation that connection was successful, along with a series of numbers that represent the communication delay in milliseconds (ms).

Q. 2. How can we find the IP Address of a Web Site ?

Ans. To find the IP address of rediffmail.com, type command in commnad prompt

```
ping rediffmail.com
```

It will return a result similar to the following that contains the IP address :

Pinging rediffmail.com [**202.137.235.12**] with 32 bytes of data : . . .

Note that many larger Web sites do not return replies to ping commands, but the site IP address can usually still be obtained. The ping method will fail if the Web site is temporarily "down" of if the computer used to perform the ping is not connected to the Internet.

Q. 3. What is ping utility ?

Ans. PING stands Packet Internet Gopher. This is a utility for ensuring connectivity between computers. ICMP protocol works behind this utility. Under it, sending node sends packets to destination node and reply is received if there is proper communication between two.

Q. 4. What are the advantages and disadvantages of three types of routing table ?

Ans. The three types of routing tables are fixed, dynamic, and fixed central. The fixed table must be manually modified every time there is a change. A dynamic table changes its information based on network traffic, reducing the amount of manual maintenance. A fixed central table lets a manager modify only one table, which is then read by other devices. The fixed central table reduces the need to update each machine's table, as with the fixed table. Usually a dynamic table causes the fewest problems for a network administrator, although the table's contents can change without the administrator being aware of the change.

REVIEW QUESTIONS

1. What is IPv4.
2. Compare IPv4 frame format with IPv6.
3. Explain following commands :
   (i) TRACERT
   (ii) ROUTE
   (iii) PING
   (iv) IPCONFIG
4. Explain frame format of TCP header.
5. Explain frame format of UDP header.
6. Compare TCP and UDP datagram.
7. What is FINGER protocol.
8. What is Ping Test ?

❑❑❑

ping www.yahoo.com

1. What is TCP?
2. [illegible]

   (i) [illegible]
   (ii) ROUTE
   (iii) PING
   (iv) [illegible]

6. [illegible] UDP datagram.
7. [illegible] protocol.
8. [illegible]

UNIT-II

4. SOCKET INTRODUCTION
5. ECHO SERVICE

Socket Introduction

Chapter 4

INSIDE THIS CHAPTER

4.1. INTRODUCTION

In computer networking of connection-based communication port is like a medium through which, an application establish a connection with another application by binding a socket by a port number. Addressing the information and the port number, accompanied the data transfer over the network.

The Ports are used by TCP and UDP to deliver the data to the right application, are identified by a 16-bit number present in the header of a data packet. Ports are typically used to map data to a particular process running on a client. If we consider a letter (data packet) sent to a particular apartment (IP) with house no. (port no), at this time the port no. is the most important part for the delivery of the letter. In order for the delivery to work, the sender needs to include an house number along with the address to ensure the letter gets to the right destination.

> **DO YOU KNOW**
>
> **TCP and UDP ports are 16 bit number.**

The TCP and UDP protocols use ports to map incoming data to a particular process running on a computer.

- Port is represented by a positive (16-bit) integer value.
- Some ports have been reserved to support common/well known services:

(i) ftp	21/tcp
(ii) telnet	23/tcp
(iii) smtp	25/tcp
(iv) login	513/tcp

User level process/services generally use port number value >= 1024

4.1.1. Types of Port

1. Well known port(0 to 1023)-They are contriolled by IANA
2. Registered Port(1024-49159)
3. Dynamic port(49152-65535)

If we consider the client-server architecture, a server application binds a socket to a specific port number in connection-based communication. It registered the server with the system where all the data destined for that port.

Now we are aware of the importance of the port number. In the same order there are some ports which are predefine and called reserved ports. Some of them are given below :

DO YOU KNOW

Port number permitt unique identification of several similtaneous process using TCP/UDP.

Reserved port numbers.

Service	Port number
echo	7
daytime	13
ftp	21
telnet	23
smtp	25
finger	79
http	80
pop3	110

DO YOU KNOW

If we consider the range of the port numbers, there are 0 to 65,535 ports available.

TIPS

The port numbers ranging from 0 - 1023 are reserved ports or we can say that are restricted ports. All the 0 to 1023 ports are reserved for use by well-known services such as FTP, telnet and http and other system services. These ports are called well-known ports.

4.2. SOCKET [MDU Dec. 2008; May 2006]

When a computer program needs to connect to a local or wide area network such as the Internet, it uses a software component called a socket. The socket opens the network connection for the program, allowing data to be read and written over the network. sockets are software, not hardware.

DO YOU KNOW

Socket is software.

Socket make it easy for software developers to create network-enabled programs. Instead of constructing network connections from scratch for each program they write, developers can just include sockets in their programs. The sockets allow the programs to use the operating system's built-in commands to handle networking functions. Because they are used for a number of different network protocols (i.e. HTTP, FTP, telnet, and e-mail), many sockets can be open at one time.

> **DO YOU KNOW**
>
> Sockets are a key part of Unix and Windows-based operating systems.

- Sockets provide an interface for programming networks at the transport layer.
- Network communication using Sockets is very much similar to performing file I/O
- In fact, socket handle is treated like file handle.
- The streams used in file I/O operation are also applicable to socket-based I/O
- Socket-based communication is programming language independent.
- That means, a socket program written in Java language can also communicate to a program written in Java or non-Java socket program.

> **DO YOU KNOW**
>
> A socket address is the combination of an IP address (the location of the computer) and a port (which is mapped to the application program process) into a single identity.

Socket Communication

A server (program) runs on a specific computer and has a socket that is bound to a specific port. The server waits and listens to the socket for a client to make a connection request.

If everything goes well, the server accepts the connection. Upon acceptance, the server gets a new socket bounds to a different port. It needs a new socket (consequently a different port number) so that it can continue to listen to the original socket for connection requests while serving the connected client.

4.3. TYPES OF SOCKET

A *socket* is one endpoint of a two-way communication link between two programs running on the network." To put it differently, it is through sockets that applications access the network and transmit data. The types of sockets are as varied as the purposes and platforms of applications. There are three types of sockets:

1. Unix Domain Sockets
2. Internet Domain Sockets /Network socket
   (a) TCP socket
   (b) UDP socket
   (c) Raw IP socket
3. NS Domain Sockets

Of these only Internet Domain Sockets are supported across all platforms. So to maintain the cross-platform characteristic intact, Java supports only Internet Domain Sockets.

Internet Domain Socket

"An Internet socket (or commonly, a socket or network socket), is a communication end-point unique to a machine communicating on an Internet Protocol-based network, such as the Internet." All applications communicating through the Internet use a network socket. The feature that distinguishes a network sockets from other sockets is the protocols that it supports. The supported protocols are:

1. TCP
2. UDP
3. Raw IP

TCP is one of the core protocols of the Internet protocol suite. The protocol guarantees reliable and in-order (correct order of packets) delivery of data from sender to receiver.

DO YOU KNOW

Internet socket are of three types : (i) Data gram socket (connection less uses UDP), (ii) Stream socket (connection oriented using TCP), (iii) Raw socket.

`it's reliable.`

The second aspect of TCP is that it is connection oriented. That means TCP requires that a connection be made between the sender and receiver before data is sent.

DO YOU KNOW

The socket associated with TCP is known as the Stream Socket.

UDP, like TCP, is one of the core protocols of the IP suite. However, unlike TCP, it neither guarantees in-order delivery of data nor does it requires a connection to be established for sending the data. UDP is

`an unreliable and`
`connectionless protocol.`

DO YOU KNOW

Sockets associated with UDP are known as Datagram Sockets.

Raw IP is a non-formatted protocol, unlike TCP and UDP. It works at network and transport layers. A socket associated with Raw IP is known as a Raw Socket. UDP and TCP sockets just receive the payload or the data, whereas Raw Sockets receive the header info of the packet along with the data. The downside of Raw Sockets is that they are tightly coupled with the implementation provided by the underlying host operating system.

TIPS

An Internet domain socket is characterized by a unique combination of the following:

- **Local socket address : Local IP address and port number**
- **Remote socket address : Only for established TCP sockets. As discussed in the Client-Server section below, this is necessary since a TCP server may serve several clients concurrently. The server creates one socket for each client, and these sockets share the same local socket address.**
- **Protocol : A transport protocol (e.g., TCP, UDP), raw IP, or others. TCP port 53 and UDP port 53 are consequently different, distinct sockets.**

Difference Between Port and Socket

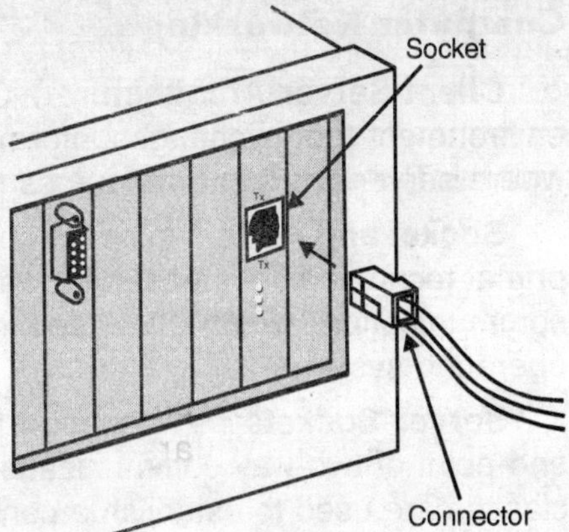

Fig. 4.1.

In common language we can say that the socket is one of the most primal technologies of computer networking. Sockets are just like an end-point of two-way communication, which allow applications to communicate using network hardware and operating systems.

However in case of java. Socket classes are used to establish a connection between client program and a server program. In java there is a **java.net** package, which provides two types of classes- first is ordinary socket, which implement the client side connection and second is server socket, which implement the server side connection.

The main purpose of the server socket is to listen an incoming connection request and ordinary socket is used to ask to server for the connection. Once a connection between client and server established, both the connected socket can communicate with each other.

In other hand we can consider the work of port in connection-based communication is like a medium through which, an application establish a connection with another application by binding a socket by a port number. Addressing the information and the port no., accompanied the data transfer over the network. The Ports are used by TCP and UDP to deliver the data to the right application, are identified by a 16-bit number.

It will take effect of registering the application with the system to receive all data bound for that port. There is a limitation for the port that no port can be bound by two applications at the same time.

4.4. SOCKETS AND JAVA SOCKET CLASSES

The Java platform is extremely preferable to write an application program require to communicate with the resources on network. Java, mainly focuses on the networking relating the description of the networking capabilities of the Java platform and second one is describes a brief summary of networking in a very simple manner that how to use URLs, sockets, and datagrams.

In Java, there is a java.net package provides the network support. All the classes for making a network program are define in the java.net package. **Through TCP we can communicate over the network.**

- A socket is an endpoint of a two-way communication link between two programs running on the network.
- A socket is bound to a port number so that the TCP layer can identify the application that data destined to be sent.
- Java's .net package provides two classes:
  - (i) Socket – for implementing a client.
  - (ii) ServerSocket – for implementing a server.

4.5. JAVA NETWORKING

Computer Networking

Client-Server Architecture : Client-server architecture can be considered as a network environment that exchanges information between a server machine and a client machine where server has some resources that can be shared by different clients.

Socket and Ports : In common language we can say that the socket is one of the most primal technologies of computer networking. Sockets are just like an end-point of two-way communication, which allow applications to communicate using network hardware and operating systems.

Server Sockets : In common language we can say that the sockets are just like an end-point of two-way communication link over the network between two programs. Socket classes are used to establish a connection between client program and a server program.

URL in Term of Java Network Programming : A URL (Uniform Resource Locator) is the address of a resource on the Internet. In java network programming we can use URLs to connect and retrieve information over the Internet.

Datagrams in Network Environment : In a network environment the client and the server communicate with each-other by reliable channel like TCP socket which have dedicated point-to-point channel between client and server. All data sent over the channel is received and sent in the same order.

Networking in Java : java.net package provides the support for networking in java with a generic style. All the java classes for developing a network program are defined in the java.net package.

Java Networking Example

1. **Find Your Host Name/IP Address :** In this section you will learn about the **getLocalHost()** method to print the Host name as well as the IP Address of the local system. For do for the same we have to call **InetAddress** class then we need to create a object, in which we store the local host information.
2. **Find Your Host Name :** Here we are going to explore a method to retrieve the host name of the local system in a very simple example. Here we are just call the InetAddress class and after made a object of that class we call a **getLocalHost()** method and pass it in the object.
3. **Getting list of Local Interfaces on a machine :** Here we are going to explain the method to find out the total no of list of local interfaces available on a machine. Here we are give a complete example named URLDemo.java.
4. **Getting the Local port :** In this section you will learn that how a user can access the Local port. Here we are going to make it easier to understand the step be step process by the complete example on this topic.
5. **Find the Host name in reverse of given IP address :** In this section you will learn to retrieve the information about the local host name using the **getHostName()** method. Here is an example that provides the usage of the **getHostName()** method in more detail.

6. **How to retrieve URL information :** Here we are going to explain a method for retrieve all the network information for a given URL. In the given below example we use the URL class and make a object.
7. **Convert URI to URL :** Here we are going to explain the method to change Uniform Resource Identifier (URI) reference to uniform resource locator (URL). Here we are give a complete example named "ConvertURItoURL.java".
8. **To retrieve the IP address from Host Name, vice-versa :** Here we are going to explain the method to find out the IP address from host name and to vice verse. Here we are give a complete example named **HostLookup.java**. In which we call a InetAddress and make a object and pass the input value in it.
9. **Construct a DatagramPacket to receive data :** In this section we provide a complete code of a example based on the method. **Datagram Packet (buffer, buffer.length)** for constructs a DatagramPacket for receiving packets of length **length** in more generic way.
10. **Construct a DatagramSocket on an unspecified port :** In this section we are going to explain the process to construct a **DatagramPacket** object in more generic way. Here we provide a complete example based on the method for creating the **DatagramSocket** object via **DatagramSocket().**
11. **Return the MIME Header :** In this section you will learn about the method to retrieve the MIME header. MIME is stands for Multipurpose Internet Mail Extensions which is an Internet Standard that extends the format of e-mail to support. A container for MimeHeader objects, which represent the MIME headers that is present in a MIME part of data.
12. **Open a URLConnection to specific website address :** Here we are going to establish a connection to a specific web address via a complete example. In the example we create a class getURLConnection and initialize a variable of url to URL and ucon of URLConnection.
13. **Print the URL of a URLConnection :** In this section we are going to describe, how to retrieve the value of the URL assign to the url object. Here is the complete code of the program in which first of all we establish a connection then display it.
14. **Getting Image from a URL :** This example shows how to get an image from the given URL.
15. **URLInformation :** Here we are going to explain the method to find out the URL information. This program defines the IOException for the exception handling. This exception is thrown to indicate an I/O problem.
16. **GetHTTPHeader :** In this section, you will learn how to get **content-length, content-type**, and **last-modify date** of a file. Here we provide a complete example that uses **getHeaderFieldKey()** method.
17. **EncoderTest :** This is a simple program of java network and supports the **java.net** package. Here, we have defined the class named **EncoderTest.java**. This class contains static methods for converting a String to the application-urlencoded MIME format.
18. **EchoClientSocket :** In this section, we are going to explain the method to find out the hostname port number of a local machine. This program defines the IOException for the exception handling.

19. **Client Socket Information :** In this section, you will learn how to get client socket information. This is a very simple program of java network. Here, we have used the class named **ClientSocketInformation.java** that implement the constructor of the **Socket** class passing two arguments as **hostName** and **TIME_PORT**.
20. **GetKeepAlive :** In this section, you will show the used of several method for different output massage. Here, we provide a complete example based on the method for creating the **Socket()** method.
21. **ReadIPAddress :** Here, we are going to explore a method to retrieve the IPAddress of the local system in a very simple example. For do for the same we have to call **InetAddress** class then we need to create a object, in which we store the local host information.
22. **Local Port Scanner :** This is a simple program of java network. Here, we are going to define class named "**LocalPortScanner**" in which we use ServerSocket class of **java.net** package which takes local port number as argument.
23. **Low port Scanner :** In this section, you will learn how to get local port number of the local machine. Here, we define the name of program **LowPortScanner**. In this we check the argument and create a socket object.
24. **URL Connection Reader :** In this section, you will learn about the url connection reader. Here, first of all we are going to define class named "**URLConnectionReader**". After that we make a URL object passing a url "**http://www.javajazzup.com**" as string in its constructor.
25. **URL Reader File :** This is a simple program of java network. In this section, you will learn how to read HTML file. Here, we create a URL object and pass a url. After that, we call the URL's **openStream()** method to get a stream from which the contents of the URL can be read.
26. **RMI Client And RMI Server Implementation :** The RMI application comprises of the two separate programs, a server and a client. A typical server program creates some remote objects, makes references to these objects accessible, and waits for clients to invoke methods on these objects.

4.6. IMPLEMENTING A SERVER

1. Open the Server Socket:
```
ServerSocket server;
DataOutputStream os;
DataInputStream is;
                              server = new ServerSocket( PORT );
```
2. Wait for the Client Request:
```
Socket client = server.accept();
```
3. Create I/O streams for communicating to the client
```
is = new DataInputStream( client.getInputStream() );
os = new DataOutputStream( client.getOutputStream() );
```
4. Perform communication with client
```
Receive from client: String line = is.readLine();
Send to client: os.writeBytes("Hello\n");
```

5. Close sockets: client.close();

For Multithreaded Server:

```
while(true) {
```

(i) wait for client requests (step 2 above)

(ii) create a thread with "client" socket as parameter (the thread creates streams (as in step (3) and does communication as stated in (4). Remove thread once service is provided.

```
}
```

Implementing a Client

1. Create a Socket Object:

```
client = new Socket( server, port_id );
```

2. Create I/O streams for communicating with the server.
   is = new DataInputStream(client.getInputStream());
   os = new DataOutputStream(client.getOutputStream());
3. Perform I/O or communication with the server:
   (i) Receive data from the server:

   ```
   String line = is.readLine();
   ```

   (ii) Send data to the server:

   ```
   os.writeBytes("Hello\n");
   ```

4. Close the socket when done:

   ```
   client.close();
   ```

A Simple Server

```
// SimpleServer.java: a simple server program
import java.net.*;
import java.io.*;
public class SimpleServer {
  public static void main(String args[]) throws IOException {
    // Register service on port 1234
    ServerSocket s = new ServerSocket(1234);
    Socket s1=s.accept(); // Wait and accept a connection
   // Get a communication stream associated with the socket
    OutputStream s1out = s1.getOutputStream();
    DataOutputStream dos = new DataOutputStream (s1out);
```

```
        // Send a string!
        dos.writeUTF("Hi there");
        // Close the connection, but not the server socket
        dos.close();
        s1out.close();
        s1.close();
    }
  }
```

A Simple Client

```
  // SimpleClient.java: a simple client program
  import java.net.*;
  import java.io.*;
  public class SimpleClient {
    public static void main(String args[]) throws IOException {
        // Open your connection to a server, at port 1234
        Socket s1 = new Socket("mundroo.cs.mu.oz.au",1234);
        // Get an input file handle from the socket and read
               the input
        InputStream s1In = s1.getInputStream();
        DataInputStream dis = new DataInputStream(s1In);
        String st = new String (dis.readUTF());
        System.out.println(st);
        // When done, just close the connection and exit
        dis.close();
        s1In.close();
        s1.close();
    }
  }
```

Run

Run Server on mundroo.cs.mu.oz.au

➢ [raj@mundroo] java SimpleServer &

Run Client on any machine (including mundroo):

➢ [raj@mundroo] java SimpleClient

➢ Hi there

If you run client when server is not up:

➢ [raj@mundroo] sockets [1:147] java SimpleClient

Exception in thread "main" java.net.ConnectException: Connection refused

at java.net.PlainSocketImpl.socketConnect(Native Method)
at java.net.PlainSocketImpl.doConnect(PlainSocketImpl.java:320)
at java.net.PlainSocketImpl.connectToAddress(PlainSocketImpl.java:133)
at java.net.PlainSocketImpl.connect(PlainSocketImpl.java:120)
at java.net.Socket.<init>(Socket.java:273)
at java.net.Socket.<init>(Socket.java:100)
at SimpleClient.main(SimpleClient.java:6)

4.7. SOCKET ADDRESS STRUCTURE [MDU Dec. 2008; May 2006]

Internet socket address structure is named **sockaddr_in** and defined by including <netinet/in.h> header.

> **DO YOU KNOW**
>
> **Programming socket programming in Java is much easier than doing it in other languages such as C.**

```
struct in_addr {
  in_addr_t    s_addr
  /* 32-bit IP address */
};                              /* network byte ordered */
struct sockaddr_in {
uint8_t  sin_len;              /* length of  structure (16) */
sa_family_t sin_family;        /* AF_INET */
in_port_t   sin_port;          /* 16-bit TCP or UDP port number */
                               /* network byte ordered */
struct in_addr sin_addr;       /* 32-bit IPv4 address */
                               /* network byte ordered */
  char     sin_zero[8];        /* unused */
```

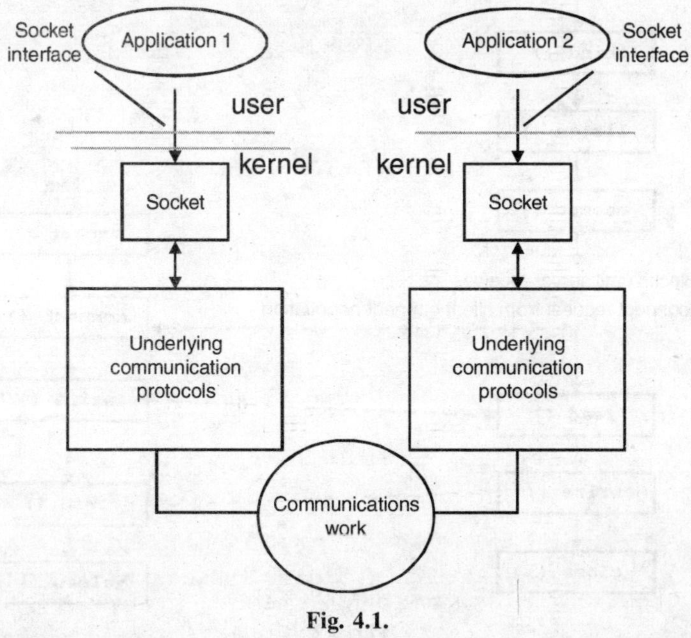

Fig. 4.1.

4.8. TCP SOCKET

A TCP socket is defined as an endpoint for communication. A socket consists of the pair

<IP Address,Port>.

A port will be defined as an integer number between 1024 and 65535. This is because all port numbers smaller than 1024 are considered *well-known* — for example, telnet uses port 23, http uses 80, ftp uses 21, and so on. On Unix machines, the file /etc/services contains a list of services provided by that machine, along with their well-known ports

A TCP connection consists of a pair of sockets. Sockets are distinguished by client and server sockets. A server listens on a port, waiting for incoming requests from clients.

Example 1. A web server listens at port 80 for incoming request from clients (web browsers). When a client wishes to make a connection with a server socket, the client is assigned a port from the local host. Suppose that client X (at IP address 146.86.3.15) wishes to browse a web page on the server 146.86.5.20. If the port

the local host assigned client X is port 12345, the connection between the client and the server is uniquely identified by the socket pair

<146.86.3.15 : 12345, 146.86.5.20 : 80>

4.8.1. TCP Clients and Servers **[MDU May 2006]**

There are two type of applications that use TCP sockets :

1. Servers
2. Clients

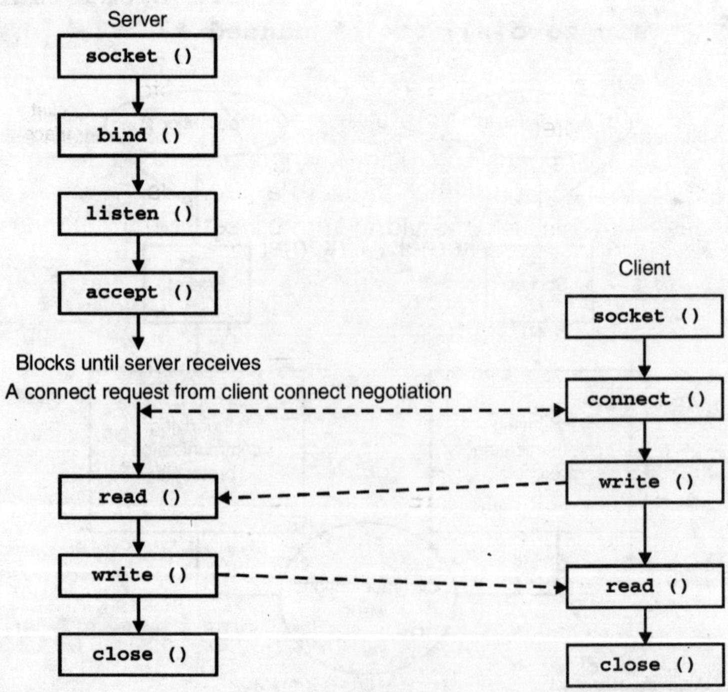

Fig. 4.2.

A TCP server listens on a well known part (or IP address and part pair) and accepts connections from TCP clients. A TCP client initiates a connection request to a TCP server in order to setup a connection with the server. A real TCP server can accept multiple connections on a socket.

The order of library calls for a TCP client is as follows:

```
socket()
connect()
send() and/or recv()
close()
```

The order of library calls for a TCP server is as follows:

```
socket()
bind()
listen()
accept()
send() and/or recv()
close()
```

4.8.2. System Calls for Elementary TCP Sockets [MDU Dec. 2006]

```
#include   <sys/types.h>
#include   <sys/socket.h>
```

> **DO YOU KNOW**
>
> System call means function call.

1. Socket ( ) Function

This function creates "an end point for communication". The return value from this function is a handle to a socket. This number is passed as a parameter to almost all of the other library calls. The type parameter can be either SOCK_STREAM (for TCP), or SOCK_DGRAM (for UDP). The protocol field is intended for specifying a specific protocol in case the network model support different types of stream and datagram models. However, TCP/IP only has one protocol for each, so this field should always be set to 0. Example to create a TCP socket:

```
int   socket   ( int  family,  int  type,  int  protocol );
```

family: specifies the protocol family {AF_INET for TCP/IP}

type: indicates communications semantics

SOCK_STREAM	**stream socket**	**TCP**
SOCK_DGRAM	**datagram socket**	**UDP**
SOCK_RAW	**raw socket**	

protocol: set to 0 except for raw sockets

returns on success: socket descriptor {a small nonnegative integer}

on error: –1

Example :

```
If  (( sd = socket (AF_INET, SOCK_STREAM, 0)) < 0)
     err_sys ("socket call error");
```

2. connect ( ) Function

This function establishes an Internet connection with the server at socket address address. The value of addrlen is always sizeof(sockaddr). The connect function blocks until either the connection is successfully established, or an error occurs. If successful, the sockfd descriptor is now ready for reading and writing. Example of a connect call:

```
int  connect (int sockfd, const struct sockaddr *servaddr,
socklen_t  addrlen);
```

sockfd: a socket descriptor returned by the socket function

*servaddr: a pointer to a socket address structure

addrlen: the size of the socket address structure

The socket address structure must contain the ***IP address*** and the ***port number*** for the connection wanted.

In TCP **connect** initiates a three-way handshake. **Connect** returns only when the connection is established or when an error occurs.

returns on success: 0

on error: -1

Example :

```
if ( connect (sd, (struct sockaddr *) &servaddr, sizeof
(servaddr)) != 0)
err_sys("connect call error");
```

3. Bind ( ) Function

Before sending and receiving data using a socket, it must first be associated with a local source port and a network interface address. The mapping of a socket to a TCP/UDP source port and IP address is called a "binding". The bind function call is used to declare the mapping between the socket, the TCP/UDP source port, and the network interface device. The prototype for bind is as follows:

> **DO YOU KNOW**
>
> **Bind assigns a local protocol address to a socket.**

```
int bind  (int  sockfd, const struct sockaddr *myaddr,
     socklen_t addrlen);
```

The first argument is a socket handle (the number returned from the socket function call). The second argument is a sockaddr structure. ck any available interface and address.

The address of the sockaddr structure is passed into the bind call, so that the socket will now be ready to communicate with remote hosts. The third parameter passed to bind is the length of the sockaddr structure.

Protocol address: a 32 bit IPv4 address and a 16 bit TCP or UDP port number.

Sockfd: a socket descriptor returned by the socket function.

*Myaddr: a pointer to a protocol-specific address.

Addrlen: the size of the socket address structure.

Servers bind their "well-known port" when they start.

returns on success: 0

on error: –1

Example :

```
If (bind (sd, (struct sockaddr *) &servaddr, sizeof
(servaddr)) != 0)
      errsys ("bind call error");
```

4. listen ( ) Function

The server listens for incoming connection requests from clients. The prototype for listen is as follows:

```
int  listen (int socket, int backlog);
```

The first argument is a socket handle (the number returned from the socket function call). The second argument specifies the maximum number of connections that can be pending on the specified socket (typically set to a large value, such as 1024).

listen is called **only** by a TCP server and performs two actions:

1. Converts an unconnected socket into a passive socket.
2. Specifies the maximum number of connections that the kernel should queue for this socket.

listen is normally called before the accept function.

returns on success: 0

on error: –1

Example :

```
If (listen (sd, 2) != 0)
      errsys ("listen call error");
```

5. accept ( ) Function

This is a blocking operation that does not return until a remote client has established a connection; when it completes, it returns a *new* socket that corresponds to this just-established connection. The prototype for accept is as follows:

```
int  accept  (int  sockfd, struct sockaddr  *cliaddr,
socklen_t  *addrlen);
```

The cliaddr argument contains the remote client's address. Note that when accept returns, the original socket that was given as argument still exists; it is used in a future invocation of accept.

accept is called by the TCP server to return the next completed connection from the front of the completed connection queue.

sockfd : This is the same socket descriptor as in listen call.

***cliaddr :** used to return the protocol address of the connected peer process (i.e., the client process).

*addrlen: {this is a value-result argument}

before the accept call: we set the integer value pointed to by *addrlen to the size of the socket address structure pointed to by cliaddr;

on return from accept call: this integer value contains the actual number of bytes stored in the socket address structure.

returns on success: a new socket descriptor

on error: –1

accept Function (cont.)

```
int    accept    (int    sockfd, struct sockaddr    *cliaddr,
socklen_t    addrlen);
```

For **accept** the first argument sockfd is the listening socket and the returned value is the connected socket.

The server will have one connected socket for each client connection accepted. When the server is finished with a client, the connected socket must be closed.

Example :

```
sfd = accept (sd, NULL, NULL);
if (sfd == -1) err_sys ("accept error");
```

6. close ( ) Function

```
int   close   (int   sockfd);
```

close marks the socket as closed and returns to the process immediately.

sockfd this socket descriptor is no longer useable.

Note : TCP will try to send any data already queued to the other end before the normal connection termination sequence.

Returns on success: 0

on error: –1

Example :

```
close (sd);
```

7. send and recv

Once a connection is established, the application processes invoke the following two operations to send and receive data:

```
int send(int socket, char * message, int msg_len, int flags);
int recv(int socket, char *buffer, int buf_len, int flags);
```

Both operations take a set of flags that control certain details of the operation.

A Client-Server Echo Program

In this once a client makes a request to the server, everything the client enters, the server echoes back.

The echo client: EchoClient.c

This client not only gets the input and output stream from the socket, but it also gets the input stream from the standard input. This allows it to read from the keyboard. Whenever the user enters a line on the keyboard, the client reads it and writes it to the socket. The client then reads from the socket, awaiting the echo back from the server.

The echo server: EchoServer.c

To compile the client and server code on Windows, make sure to add the library module Wsock32.lib to your project .

To compile the client and server code On Unix, you can use this Makefile.

```
#
# makefile for echo client and server
#

CC = gcc
LIBS = -lsocket -lnsl -lpthread

all: EchoServer EchoClient

EchoServer: EchoServer.c
	$(CC) -o EchoServer EchoServer.c $(LIBS)

EchoClient: EchoClient.c
	$(CC) -o EchoClient EchoClient.c $(LIBS)

clean:
	/bin/rm -f EchoServer EchoClient core *.o *~
```

Testing the Client-Server Echo Program

To test the echo client-server application, start the server by entering (in Unix)

```
./EchoServer <port number>
```

Now you can test the server using both telnet and the client.

(a) **Using telnet :** in a separate window, enter the command:

```
telnet localhost <port number>
```

Now, everything you enter on that window will be echoed back to you. To quit, open another terminal on tanner and kill both applications (EchoServer and telnet).

(a) **Using the echo client :** in a separate window, start the echo client by entering (in Unix)

```
./EchoClient localhost <port number>
```

Now, everything you enter on that window will be echoed back to you. Quit the client with CTRL-D (this is the end-of-file character in Unix).

4.9. ELEMENTARY UDP SOCKETS

[MDU Dec. 2007]

Following are the properties of UDP sockets :

- Connectionless
- Unreliable
- Datagram protocol
- Popular using
  DNS (the Domain Name System)
  NFS (the Network File System)
  SNMP (Simple Network Management Protocol)

> **DO YOU KNOW**
>
> **The EchoServer program can handle one single client at a time. For instance, if you try to run concurrently two echo clients (in two separate windows), only one client can talk to the server; the second client can start talking to the server only after the first client finishes its execution.**

4.9.1. Function Call for UDP Client Server **[MDU Dec. 2007; Dec. 2006]**

1. Client does not establish a connection with the server.
2. Server does not accept a connection from a client.
3. recvfrom function returns the protocol address of the client so server can send a response to correct client.

Following are the socket function for UDP client server.

Following are input function for elementary UDP client/server are

1. send to ( )
2. recv from ( )
3. socket ( )
4. bind ( )
5. close ( )

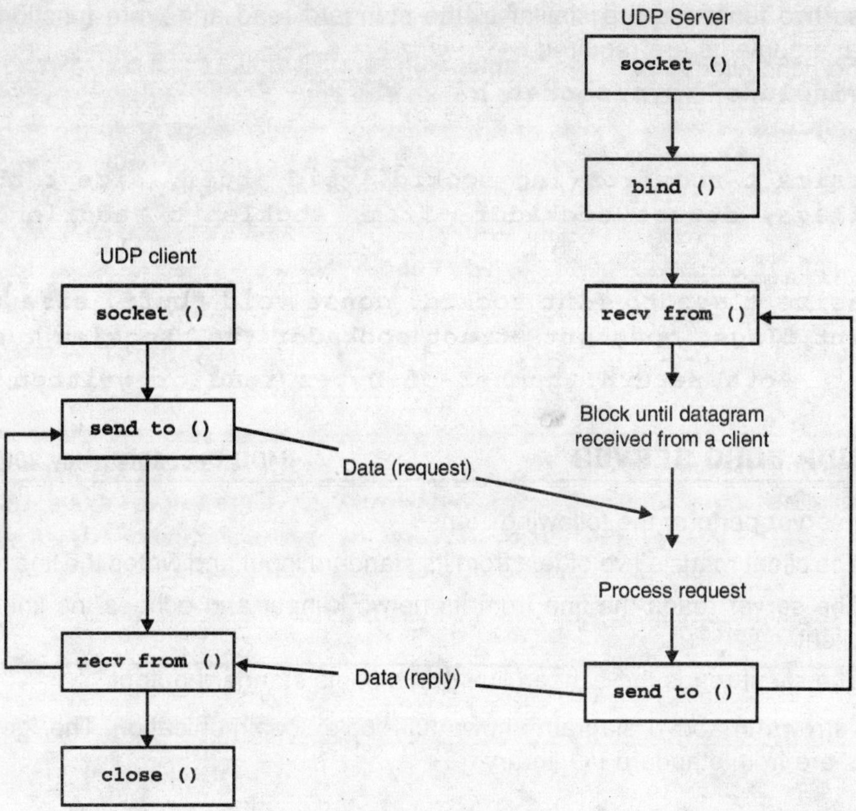

Socket functions for UDP client-server

Fig. 4.3.

4.9.2. Recvfrom and Sendto Functions

Sockfd	- socket Descriptor.
*buff	- Pointer to buffer to read or write.
Nbyutes	- Number of bytes to read or write.
flags	- Set to 0 for now

sendto	to	socket address structure containing the protocol address(IP address and port#)of where data is to be sent.
	addrlen	Size of socket address structure (an interger value).
recvfrom	from	socket address structure that is filled in by recvfrom with the protocol addres (ip address and port #) of where data came from.
	* addrlen	(pointer to integer value)

These two functions are similar to the standard read and write functions, but three additional arguments are required.

```
#include <sys/socket.h>

ssize_t recvfrom(ing sockfd, void *buff, size_t nbyes, int
flags, struct sockaddr *from, socklen_t *addrlen):

ssize_t sendto (int sockfd, const void *buff, size_t nbytes,
int flags, constant struct sockaddr *to, socklen_t addrlen);
    Both return: number of bytes read or written if OK, -
```

4.10. UDP ECHO SERVER [MDU Dec. 2008; May 2008; Dec. 2007]

An echo server perform the following steps :

1. The client roads a live of text from its standard input and writes the line to the server.
2. The server reads the line from its network input and echoes the line back to the client.
3. The client reads the echoed line prints on its standard output.

Two arrows in above diagram shows full duplex communication. The fgets and fputs functions are from standard I/O library.

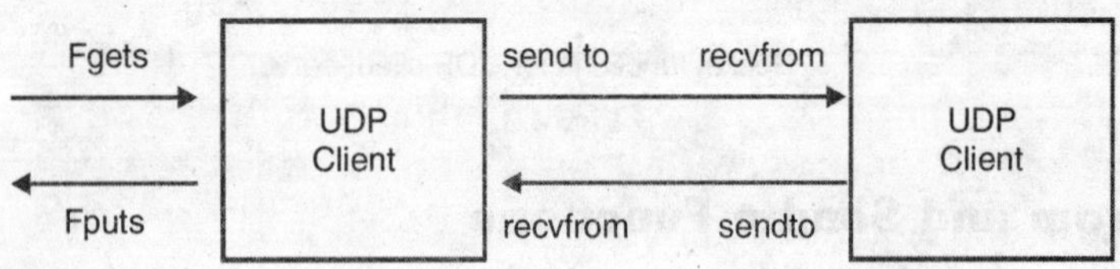

Fig. 4.4.

Server Main Function

```
#include *unp.h*
_______________________________udpcliserv/udpserv01.c
int
main(int argc, char **argv
{
   int socketfd;
   struct sockaddr_in servaddr, cliaddr

   sockfd = Socket (AF_INET, SOCK_DGRAM, 0);

   bzero (&servaddr, size of (servaddr));
```

```
9    servaddr.sin_family = AF_INET;
10   servaddr.sin_addr.s_addr = htonl (INADDR_ANY);
11   servaddr.sin_port = htons(SERV_PORT);

12   Bind (sockfd, (SA *) &servaddr, sizeof (servaddr));

13   dg_echo (sockfd, (SA*) &cliaddr, size of(cliaddr));
14   }
```

udpcliserv/udpserv01.c

Fig. 4.5. UDP Echo Server

Server main function

Lines 7-12:

Note now we use SOCK_DGRAM to mean UDP

INADDR_ANY = 0

SERV_PORT = 9877

Line 13

dg_echo does the server processing

Server dg_echo function

Line 8-12:

Loop reads next datagram and sends back.

4.10.1. UDP Echo Server : dg_echo Function

Figure 4.8 shows the dg_echo function.

lib/dg_echo.c

```
1  #include "unp.h"
2  void
3  dg_echo (int sockfd, SA *pcliaddr, socklen_t clilen)
4  (
5     int   n;
6     socklen_t len;
7     char   mesg (MAXLINE);

8     for (;  ;)  {
9        len = clilen;
10       n = Recvfrom (sockfd, mesg, MAXLINE, 0, pcliaddr, &len);
```

```
Sendto (sockfd, mesg, n, 0, pcliaddr, len);
    }
}
```

Fig. 4.6. dg_echo Function: Echo Lines on a Datagram Socket

Server dg_echo Function

This dg-echo function never terminates. This is an iterative server which is typical for UDP.

A concurrent server calls fork function, TCP is typically concurrent. Each UDP socket has a receive buffer. (Aside: size is controlled by a socket

`option SO_RCVBUF ).`

4.10.2. Difference Between TCP and UDP Client Server With Two Clients

Most TCP servers are concurrent and most UDP servers are iterative.

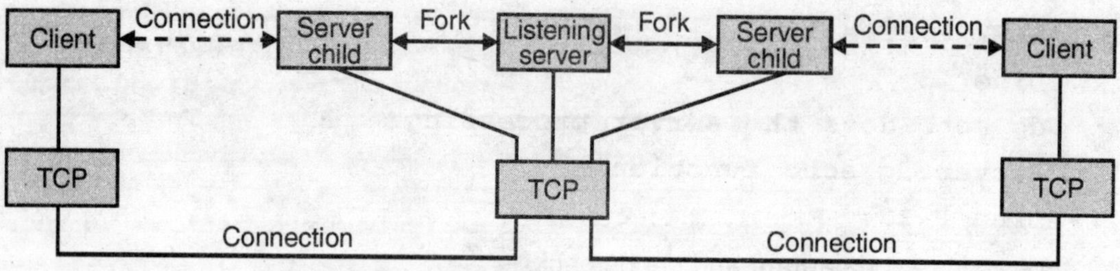

Fig. 4.7. TCP Client Server With Two Clients

UDP Client Server With Two Clients

There are two connected sockets and each of the two connected sockets on the server host has its own socket receive buffer.

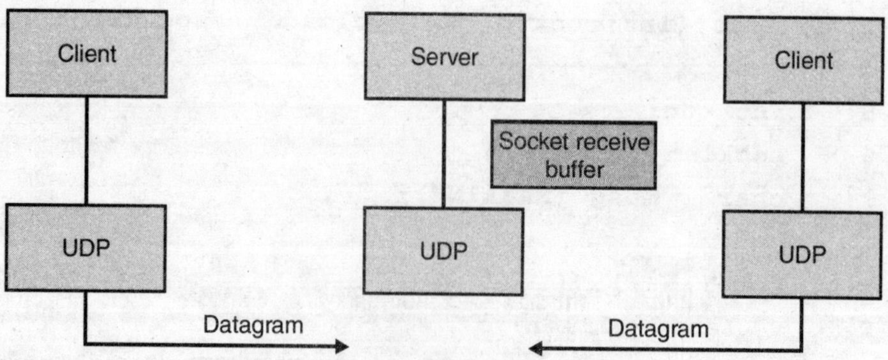

Fig. 4.8. UDP Client Server With Two Clients.

4.10.3. Problems With UDP Sockets

> **DO YOU KNOW**
>
> UDP Use a request-reply model.

Lost datagrams (client request or server reply):

➢ recvfrom blocks forever

➢ place a timeout, but don't know whether request or reply gets lost

Malicious datagrams inter-mixed with server replies:

➢ ignore any received datagrams not from that server

➢ allocate another socket address structure and compare returned address

For multi-homed server, verifying address may not work (server reply may go through another outgoing interface)

➢ solution 1: verify server domain name, instead

➢ solution 2: multi-homed UDP server creates one socket for every interface (IP addr), bind IP addresses to sockets, use *select* across all sockets

Server not running :

➢ ICMP port unreachable error (asynchronous error)

➢ Asynchronous errors not returned for UDP sockets unless the socket has been connected (reason?: considering a client sending 3 datagrams to 3 servers, recvfrom has no way to know the destination of the datagram causing the error)

➢ Recvfrom blocks forever

➢ solution: call *connect* on a UDP socket

Lack of flow control:

> **DO YOU KNOW**
>
> UDP sockets may generate asynchronous errors reported only to connected sockets.

➢ considering dg_cli in a client sendto 2000 1400-byte datagrams to the server

➢ client may overrun the server (e.g. 96% loss rate: mostly lost due to receive buffer overflow, some due to network congestion)

➢ use netstat -s to check the loss

➢ solution: use SO_RCVBUF option to enlarge buffer, use request-reply model instead of bulk transfer

4.11. SELECT() AND POLL() FUNCTION [MDU Dec. 2008; May 2008]

The *select()* and *poll()* methods are powerful tool for multiplexing network sockets. Specifically, these methods will indicate when a procedure will be safe to execute on an open file descriptor without any delays. For instance, a programmer can use these calls to know when there is data to be read on a socket. By delegating responsibility to *select()* and *poll()*, you don't have to constantly check whether there is data to be read. Instead, *select()* and *poll()* can be placed in the background by the operating system and woken up when the event is satisfied or a specified timeout has elapsed.

Advantages of Using Select nd Poll

This process can significantly increase execution efficiency of a program.

select() : This function Check if sockets descriptors are ready to read/write

Syntax:

```
int select(int n, fd_set *readfds, fd_set *writefds, fd_set
*exceptfds, struct timeval *timeout);

FD_SET(int fd, fd_set *set);
FD_CLR(int fd, fd_set *set);
FD_ISSET(int fd, fd_set *set);
FD_ZERO(fd_set *set);
```

The helper macros do the following:

```
FD_SET(int fd, fd_set *set);   Add fd to the set.
FD_CLR(int fd, fd_set *set);   Remove fd from the set.

FD_ISSET(int fd, fd_set *set); Return true if fd is in the set.
FD_ZERO(fd_set *set); Clear all entries from the set.
```

Select() function Returns the number of descriptors in the set on success,

0 if the timeout was reached,

or –1 on error (and **errno** will be set accordingly.) Also, the sets are modified to show which sockets are ready.

poll() : This function Test for events on multiple sockets simultaneously

Syntax

```
int poll(struct pollfd *ufds, unsigned int nfds, int timeout);
```

This function is very similar to **select()** in that they both watch sets of file descriptors for events, such as incoming data ready to **recv()**, socket ready to **send()** data to, out-of-band data ready to **recv()**, errors, etc.

The basic idea is that you pass an array of *nfds* struct pollfds in *ufds*, along with a timeout in milliseconds (1000 milliseconds in a second.) The *timeout* can be negative if you want to wait forever. If no event happens on any of the socket descriptors by the timeout, **poll()** will return.

Each element in the array of struct pollfds represents one socket descriptor, and contains the following fields:

```
struct pollfd {
    int fd;           // the socket descriptor
    short events;     // bitmap of events we're interested in
    short revents;    // when poll() returns, bitmap of events
                                  that occurred
};
```

Before calling **poll()**, load *fd* with the socket descriptor and then construct the *events* field by bitwise-ORing the following macros:

POLLIN **Alert me when data is ready to** recv() **on this socket.**

POLLOUT **Alert me when I can** send() **data to this socket without blocking.**

POLLPRI **Alert me when out-of-band data is ready to** recv() **on this socket.**

Once the **poll()** call returns, the *revents* field will be constructed as a bitwise-OR of the above fields, telling you which descriptors actually have had that event occur. Additionally, these other fields might be present:

POLLERR	**An error has occurred on this socket.**
POLLHUP	**The remote side of the connection hung up.**
POLLNVAL	**Something was wrong with the socket descriptor** ***fd*****—maybe it's uninitialized?**

Poll() function Returns the number of elements in the *ufds* array that have had event occur on them; this can be zero if the timeout occurred. Also returns -1 on error (and **errno** will be set accordingly.

4.12. ADDRESS CONVERSION [MDU Dec. 2008; Dec. 2008]

Various functions are used for address conversion

1. inet_ntoa( ),
2. inet_aton( ),
3. inet_addr( )
4. inet_ntop( ),
5. inet_pton( )

1. The function **inet_ntoa()** converts a network address in a struct in_addr to a dots-and-numbers format string. The "n" in "ntoa" stands for network, and the "a" stands for ASCII for historical reasons (so it's "Network To ASCII"—the "toa" suffix has an analogous friend in the C library called **atoi()** which converts an ASCII string to an integer.)

inet_ntoa() returns the dots-and-numbers string in a static buffer that is overwritten with each call to the function.

2. The function **inet_aton()** is the opposite, converting from a dots-and-numbers string into a in_addr_**inet_aton()** returns non-zero if the address is a valid one, and it returns zero if the address is invalid.

3. The function **inet_addr()** is an older function that does basically the same thing as **inet_aton()**. It's theoretically deprecated, but you'll see it a lot and the police won't come get you if you use it.

   inet_addr() returns the address as an in_addr_t, or -1 if there's an error. (That is the same result as if you tried to convert the string "255.255.255.255", which is a valid IP address. This is why **inet_aton()** is better.)

4. inet_ntop()-It Convert IP addresses to human-readable form and back.

   inet_ntop() returns the *dst* parameter on success, or NULL on failure (and *errno* is set).

5. inet_pton()It Convert IP addresses to human-readable form and back **inet_pton()** returns 1 on success. It returns -1 if there was an error (*errno* is set), or 0 if the input isn't a valid IP address

These functions are for dealing with human-readable IP addresses and converting them to their binary representation for use with various functions and system calls. The "n" stands for "network", and "p" for "presentation". Or "text presentation".

4.13. OTHER SYSTEM CALL

1. Accept()

Accept an incoming connection on a listening socket

Syntax

```
int accept(int s, struct sockaddr *addr, socklen_t *addrlen);
```

accept() returns the newly connected socket descriptor, or -1 on error, with *errno* set appropriately.

2. bind()

Associate a socket with an IP address and port number.

Syntax

```
int bind(int sockfd, struct sockaddr *my_addr, socklen_t
addrlen);
```

When a remote machine wants to connect to your server program, it needs two pieces of information: the IP address and the port number. The **bind()** call allows you to do just that.

First, you call **getaddrinfo()** to load up a **struct sockaddr** with the destination address and port information. Then you call **socket()** to get a socket descriptor, and then you pass the socket and address into **bind()**, and the IP address and port are bound to the socket!

Returns zero on success, or -1 on error

3. Connect()

Connect a socket to a server

Syntax

```
int connect(int sockfd, const struct sockaddr *serv_addr,
socklen_t addrlen);
```

Once you've built a socket descriptor with the **socket()** call, you can **connect()** that socket to a remote server using the well-named **connect()** system call

Returns zero on success, or -1 on error (and **errno** will be set accordingly.)

4. close()

Close a socket descriptor

Syntax:

```
int close(int s);
```

Returns zero on success, or –1 on error (and **errno** will be set accordingly.

> **DO YOU KNOW**
>
> **Windows users will use the function closesocket(), not close() for closing the socket**

5. gethostname()

Returns the name of the system

Syntax:

```
int gethostname(char *name, size_t len);
```

Returns zero on success, or -1 on error (and **errno** will be set accordingly.)

6. gethostbyname()

get IP address for a host name

syntax:

```
struct hostent *gethostbyname(const char *name);
```

Returns a pointer to a resultant struct hostent or success, or NULL on error.

7. gethostbyaddr()

get host name for IP address. It is reverse of gethostbyname() function

```
struct hostent *gethostbyaddr(const char *addr, int len,
int type);
```

Returns a pointer to a resultant struct hostent or success, or NULL on error.

8. getnameinfo()

Look up the host name and service name information for a given struct sockaddr

Syntax:

```
int getnameinfo(const struct sockaddr *sa, socklen_t salen,
char *host, size_t hostlen, char *serv, size_t servlen,
int flags);
```

This function is the opposite of **getaddrinfo()**, that is, this function takes an already loaded struct sockaddr and does a name and service name lookup on it. It replaces the old **gethostbyaddr()** and **getservbyport()** functions

Returns zero on success, or non-zero on error. If the return value is non-zero, it can be passed to **gai_strerror()** to get a human-readable string. See **getaddrinfo** for more information.

9. getpeername()

Return address info about the remote side of the connection

Syntax:

```
int getpeername(int s, struct sockaddr *addr, socklen_t
*len);
```

Returns zero on success, or -1 on error (and **errno** will be set accordingly.)

10. *errno*

Holds the error code for the last system call

Syntax

```
int errno;
```

The value of the variable is the latest error to have transpired, which might be the code for "success" if the last action succeeded.

11. fcntl()

Control socket descriptors

Syntax:

```
int fcntl(int s, int cmd, long arg);
```

This function is typically used to do file locking and other file-oriented stuff,

Returns zero on success, or –1 on error (and **errno** will be set accordingly.)

12. htons(), htonl(), ntohs(), ntohl()

Convert multi-byte integer types from host byte order to network byte order

Syntax:

```
uint32_t htonl(uint32_t hostlong);
uint16_t htons(uint16_t hostshort);
uint32_t ntohl(uint32_t netlong);
uint16_t ntohs(uint16_t netshort);

htons() host to network short
htonl() host to network long
ntohs() network to host short
ntohl() network to host long
```

Each function returns the converted value.

13. listen()

Tell a socket to listen for incoming connections

Syntax:

```
int listen(int s, int backlog);
```

Before calling **listen()**, your server should call **bind()** to attach itself to a specific port number. That port number (on the server's IP address) will be the one that clients connect to.

Returns zero on success, or -1 on error (and **errno** will be set accordingly.)

14. perror(), strerror()

Print an error as a human-readable string

```
S void perror(const char *s);
char *strerror(int errnum); syntax
```

> **DO YOU KNOW**
>
> **The function strerror() is very similar to perror(), except it returns a pointer to the error message string for a given value.**

strerror() returns a pointer to the error message string

15. recv(), recvfrom()

Receive data on a socket

Syntax:

```
ssize_t recv(int s, void *buf, size_t len, int flags);

ssize_t recvfrom(int s, void *buf, size_t len, int flags,
struct sockaddr *from, socklen_t *fromlen);
```

Returns the number of bytes actually received (which might be less than you requested in the *len* parameter), or -1 on error. If the remote side has closed the connection, **recv()** will return 0. This is the normal method for determining if the remote side has closed the connection.

16. setsockopt(), getsockopt()

Set various options for a socket

Syntax:

```
int getsockopt(int s, int level, int optname, void *optval,
                 socklen_t *optlen);
int setsockopt(int s, int level, int optname, const void
*optval,          socklen_t optlen);
```

Returns zero on success, or -1 on error (and **errno** will be set accordingly.)

17. send(), sendto()

Send data out over a socket.

```
ssize_t send(int s, const void *buf, size_t len, int flags);

ssize_t sendto(int s, const void *buf, size_t len,
          int flags, const struct sockaddr *to,
                socklen_t tolen);
```

These functions send data to a socket. **send()** is used for TCP SOCK_STREAM connected sockets, and **sendto()** is used for UDP SOCK_DGRAM unconnected datagram sockets.

With the unconnected sockets, you must specify the destination of a packet each time you send one, and that's why the last parameters of **sendto()** define where the packet is going.

With both **send()** and **sendto()**, the parameter *s* is the socket, *buf* is a pointer to the data you want to send, *len* is the number of bytes you want to send, and *flags* allows you to specify more information about how the data is to be sent.

Returns the number of bytes actually sent, or –1 on error

18. shutdown()

Stop further sends and receives on a socket

Syntax:

```
int shutdown(int s, int how);
```

Returns zero on success, or –1 on error

DO YOU KNOW

When you close() a socket descriptor, it closes both sides of the socket for reading and writing, and frees the socket descriptor. If you just want to close one side or the other, you can use this shutdown() call.

19. Socket()

It Allocate a socket descriptor

Syntax

```
int socket(int domain, int type, int protocol);
```

The new socket descriptor to be used in subsequent calls, or -1 on error

HIGHLIGHTS

1. Port address is of 16 bits.
2. Port are :
   - ➢ Well known port (0 to 1023)
   - ➢ Registered port (1024-49159)
   - ➢ Dynamic port (49152-65535)
3. The socket opens the network connection for the program, allowing data to be read and written over the network.
4. Socket are software.
5. socket address = port address + IP address
6. Types of socket :
   - ➢ Unix domain sockets
   - ➢ Internet domain sockets
     - ● TCP socket
     - ● UDP socket
     - ● Raw IP socket
   - ➢ NS domain socket
7. TCP socket function call consists of following function

TCP server function	TCP client function
➢ socket ( )	socket ( )
➢ bind ( )	connect ( )
➢ listen ( )	write ( )
➢ accept ( )	read ( )
➢ read ( )	close ( )
➢ write ( )	
➢ close ( )	

8. UDP socket call consists of following function call socket.

UDP client function	UDP client function
➢ socket ( )	socket ( )
➢ send to ( )	bind ( )
➢ recv from ( )	recvfrom ( )
➢ close ( )	send to ( )

9. Following are problems with UDP sockets :
   - Last datagrams
   - Malicious datagrams
   - Server not running
   - Lack of flow control
10. Select ( ) function check if socket descriptors are ready to read/write.
11. Poll function test for events on multiple socket simultaneously.
12. Various functions are used for address conversion.
   - inet-ntoa( ) → IP address → dots and number string
   - inet_aton ( ) → dots and number → IP address
   - inet_addr ( ) → same as inet_aton
   - inet_addr ( ) → same as inet_aton
   - inet_ntop ( ) IP address → Human readableform
   - inet_pton( ) IP address → Human readleform

QUESTIONS WITH ANSWERS

Q. 1. What is IANA?

Ans. Internet Assigned Number Authority. IANA divides the port address

Q. 2. What is mean by well known port?

Ans. In well known port, port no and the port address are 16 bits. The range is from 0 to 1023. It is assigned and controlled by IAN. It is used in server process.

Q. 3. Compare physical, logical and port address.

Q. 4. Write a program to print its own process ID.

Ans.

```
#include<stdio.h>
#include<stdio.h>
int main(void)
{
pid_t pid;
pid=getpid();
printf("My pid = %d",pid);
}
```

Q. 5. Write a program to print the process ID of parent and child.

Ans.

```
#include<stdlib.h>
#include<stdio.h>
int main(void)
{
```

```
pid_t pid;
pid=fork();
if(pid==0)
printf("Child pid = %d\n",pid);
pid=getppid();
printf("My parent pid = %d\n",pid);
return 0;
}
```

Q. 6. Define socket. Give structure of socket.

Ans. In client server model two application programs one running on local system(client) and another running on remote system(srever) need to communicate with one another. The communication structure that we need in socket programming is socket.

Q. 7. Define the types of socket.

1. Stream socket(SOCK_STREAM)
2. Datagram socket(SOCK_DGRAM)
3. Raw socket(SOCK_RAW)

Q. 8. Explain the fields present in sockaddr_in structure.

```
struct sockaddr_in
{
uint8_t sin_len;                /*length of the socket*/
sa_family_t sin_family;         /*AF_INET*/
in_port_t sin_port;             /*16 bit port addr*/
struct in_addr sin_addr;        /*32 IPV4 addr*/
char sin_zero[8];               /*unused field*/
}
struct int_addr
{
in_addr_t s_addr;               /*32 bit IPV4 addr*/
};
```

```
Q9.Specify the fields in sockaddr_in6.
ans
struct int_addr
{
uint8_t sa_addr[16];            /*128 bit IPV6 addr*/
};
#define SIN6_LEN
```

```
struct sockaddr_in6
{
uint8_t sin6_len;            /*length of the socket*/
sa_family_t sin6_family;     /*AF_INET6*/
in_port_t sin6_port;         /*16 bit addr*/
uint32_t sin6_flowinto;      /*flow control*/
struct in6_addr sin6_addr;       /*128 IPV6 socket addr*/
};
```

Q. 9. What are the data types used in client server program.

Ans. The data types used in client server program are

Q. 10. What is network byte order?

Ans. Network byte order chooses its own byte order by the protocols used. It may be either LEBO or BEBO.

Q. 11. What is the purpose of inet_aton function?

Ans. The purpose of inet_aton function is to convert the dotted decimal format to 32 bit binary format in network byte order.

General Syntax :

```
int inet_aton(const char *strptr, struct in_addr  *addrptr);
```

Q. 12. How a socket is uniquely identified?

Ans. The socket function returns a small non-negative integer value, similar to a file descriptor. So we call this to be socket descriptor,

General Syntax :

```
int socket(int family, int type, int protocol);
```

Q. 13. What is the purpose of bind and connect?

Ans. BIND: Bind function is used to assign the local socket address to already created socket.

General Syntax :

```
int bind(int sockfd, const struct sockaddr *myaddr, socklen_t
addrlen);
```

CONNECT: Connect function is used by TCP client to establish active connection with TCP server.

General syntax :

```
int connect(int sockfd, const struct sockaddr *serveraddr,
socklen_t addrlen);
```

Q. 14. What is the need for listen and accept system calls?

Ans. LISTEN: Listen function enforce operating system that the server is ready to accept the connection through the socket.

General Syntax :

```
int listen(int sockfd, int backlog);
```

ACCEPT : Accept function is called by a TCP server to return the next completedconnection from the front of the completed connection queue.

General Syntax :

```
int accept(int sockfd, struct sockaddr cliaddr, sockaddrlen_t
addrlen)
```

Q. 15. What is the need for listen and accept system calls?

Ans. **LISTEN :** Listen function enforce operating system that the server is ready to accept the connection through the socket.

General Syntax :

```
int listen(int sockfd, int backlog);
```

ACCEPT : Accept function is called by a TCP server to return the next completed connection from the front of the completed connection queue.

General Syntax :

```
int accept(int sockfd, struct sockaddr cliaddr, sockaddrlen_t
addrlen);
```

Q. 16. Specify address conversion functions.

Ans. There are five conversion functions. They are as follows:

1. inet_aton

```
int inet_aton(const char *strptr, struct in_addr *addrptr);
```

2. inet_ntoa

```
char *inet_ntoa(struct in_addr inaddr);
```

3. inet_addr

```
in_addr_t inet_addr(const char *strptr);
```

4. inet_pton

```
int inet_pton(int family, const char *strptr, void *addrptr);
```

5. **inet_ntop**

```
const char *inet_ntop(int family, const void *addrptr, char

*strptr, size_t len);
```

Q. 17. Explain the operation of fork() system call.

Ans. Fork() causes the process to split into two identical process one retains the original process identification number(PID) and is called parent. The other called child gets a new PID.

Q. 18. What are the functions used for byte manipulation operations?

Ans. Two groups of functions are for byte manipulation

```
« b (byte)
| bzero
| bcopy
```

```
| bcm
« mem (memory)
| memset
| memcopy
```

Q. 19. What is socket abstraction?

Ans. Socket is an abstraction through which an application may send and receive data, in much the same way as an open file allows an application to read and write data to stable storage.

Q. 20. Compare various socket address structure.

```
Ipv4
SOCKADDR_IN4{}

Length          AF_INET
16 bit port#
32 bit IPV4 addr
Unused 8 bytes

IPV6
SOCKADDR_IN6{}

Length              AF_INET6
16 bit port#
32 bit FLOW LABEL
128 bit
IPV6 addr

UNIX
SOCKADDR_UN{}
Length          AF_LOCAL
Pathname(upto 104
bytes)
DATALINK
SOCKADDR_DL{}
Length AF_LINK
Interface index
Type namelen
addrlen sel len
Interface name and link
```

```
layer addr
```

Q. 21. Write short notes on sendto and recvfrom system call.

Ans. SENDTO: This function is used by a UDP process to send a message to another process running on remote machine.

General Syntax :

```
ssize_t sendto(int sockfd, const void *buf, size_t buflen,
int flags, const struct
sockaddr *toaddr, socklen_t toaddrlen);
```

RECVFROM : This function is used by a UDP process to send a reply to the sender.

General Syntax :

```
ssize_t recvfrom(int sockfd, const void *buf, size_t buflen,
int flags, struct
sockaddr *fromaddr, socklen_t fromaddrlen);
```

Q. 22. What is meant by polling?

Ans. An application that lies in a loop calling recvfrom on a non-blocking I/O model is called polling.

Q. 23. What is select function? [MDU Dec. 2008]

Ans. Select function monitor several sockets at the same time. This function return ready socket. select function allows the process to instruct the kernel to wait for any one of multiple events to occur and to wake up the process only when one or more of the events occurs or when a specified amount of time has passed.

Q. 24. Describe the general format of select function? [MDU Dec. 2008]

Ans. The general format is,

```
int select (int maxfd1, fd_set *readset, fd_set *writeset,
fd_set *exceptset,
const struct timeval *timeout);
return positive count of ready descriptors
0 if on timeout
-1 if error
```

Q. 25. Define wait forever?

Ans. Wait forever is that, it returns only when one of the specified descriptor is ready for I/O.

Q. 26. What is Donot wait at all?

Ans. Do not wait at all states that, return immediately after checking the descriptor. Timeout point to timeval structure. Timeval value must be zero

Q. 27. What are the conditions of the descriptor to become ready?

Ans. The conditions of the descriptor to become ready,

- ➢ If the socket is ready for reading
- ➢ If the socket is ready for writing
- ➢ Socket has an exception pending

Q. 28. What are the steps involved for socket has exception pending?

Ans. The steps involved for socket has exception pending are,

- There exist an out of band data for the socket.
- Socket is in the out of band mark.

Q. 29. What is the difference between shutdown and close function?

Ans. # With shutdown, we can initiate TCP's normal connection termination sequence regardness of reference count.

In this case of close function, decrement the reference count and closes the socket only if the count reaches 0.

Shutdown function tell the other end that we have finish sending eventhough that end might have to send more data's.

In this case of close function, terminates both directions of data transfer reading and writing.

Q. 30. What is poll function? **[MDU Dec. 2008]**

Ans. Poll function test for events on multiple socket simultaneously. It should include the following header file.

```
#include<poll.h>
int poll (struct pollfd *fdarray, unsigned long *nfds, int
timeout);
returns count of ready descriptors
0 on timeout
-1 on error
```

REVIEW QUESTIONS

1. Define socket and List out its types.
2. Explain socket address structure.
3. What are the four functions to pass socket address structure from kernel to the process?
4. What are elementary socket functions?
5. Explain TCP client-server.
6. Explain UDP client-server.
7. What are various TCP socket call. What is their order of execution ?
8. What are various UDP socket call.
9. What is port. Differentiate port and socket.
10. Explain following TCP socket call :
    (i) bind ( )
    (ii) Listen ( )
    (iii) Accept
11. Explain sendto ( ) and recvfrm ( ) socket call.
12. What are reserved port numbers.
13. What is echo server.
14. Explain various address conversion function.

❑❑❑

Echo Service

INSIDE THIS CHAPTER

Chapter **5**

5.1. INTRODUCTION

The **Echo Protocol** is a service in the Internet Protocol Suite defined in RFC 862. It was proposed for testing and measurement of round-trip times. In IP networks.

> **DO YOU KNOW**
>
> **Echo is small server of TCP and UDP.**

A host may connect to a server that supports the Echo Protocol using the Transmission Control Protocol (TCP) or the User Datagram Protocol (UDP) on the well-known port number 7. The server sends back a copy of the identical data it received.

On unix operating systems an echo server is built into the inetd command. The echo service is usually not enabled by default. It may be enabled by adding the following lines to the file /etc/inetd.conf and telling inetd to reload its configuration:

```
echo    stream   tcp     nowait   root     internal

echo    dgram    udp     wait     root     internal
```

> **TIPS**
>
> **Echo uses UDP and TCP port 7 and is used as a debgging tool to send any datagrams received from a source, back to that source.**

Service	Port	Protocol
echo	7	tcp
echo	7	udp
discard	9	tcp
discard	9	udp
systat	11	tcp
systat	11	tcp
daytime	13	tcp
daytime	13	udp
netstat	15	tcp
qotd	17	tcp
qotd	17	udp
chargen	19	tcp
chargen	19	udp
ftp-data	20	tcp
ftp	21	tcp
telnet	23	tcp
smtp	25	tcp
time	37	tcp
time	37	udp
rlp	39	udp
name	42	tcp
name	42	udp
whois	43	tcp
domain	53	tcp
domain	53	udp
nameserver	53	tcp
nameserver	53	udp
mtp	57	tcp
bootp	67	udp
tftp	69	udp
gopher	70	tcp
rje	77	tcp
finger	79	tcp
http	80	tcp
link	87	tcp
supdup	95	tcp

Discard

Discard is another debugging tool where data is sent to port **9** and the recipient simply throws it away. This data can either be a UDP packet or a TCP stream that is opened until closed by the sender.

DO YOU KNOW

Echo service is defined by RFC 862.

TIPS

Discard is small server of TCP and UDP.

Chargen

Chargen is short for Character Generator and is a service that generates random characters either in one UDP packet containing a random number (between 0 and 512) of characters, or a TCP session. The UDP Chargen server looks for a UDP packet on port 19 and responds with the random character packet.

With TCP Chargen, the server sends as a continuous stream of TCP packets once a connection is made, and until the session closes. The data is thrown away. Chargen is used to find the cause for dropped packets. It uses TCP/UDP port **19**. An infiltrator can create a DoS attack by spoofing an IP address and causing two devices to send random traffic to each other.

DO YOU KNOW

RFC 863 describes the Discard service.

DO YOU KNOW

The Chargen service is described by RFC 864.

TIPS

Chargen is small server of TCP and UDP.

Daytime

The Daytime protocol is another testing tool and uses port **13** for both UDP and TCP versions. On receipt of a datagram to port 13, the server in the UDP version, sends the current date and time in ASCII format in a datagram. The TCP version, on receipt of a datagram to port 13, and once the connection is established, sends the date and time in ASCII format and closes the connection.

DO YOU KNOW

The Daytime service is described by RFC 867.

TIPS

Daytime is small server of tcp only.

5.2. TCP AND UDP SMALL SERVERS

TCP and UDP small servers are servers that run in the router which are useful for diagnostics.

TCP Small Servers

The TCP small servers are:

- **Echo :** Echoes back whatever you type through the **telnet x.x.x.x echo** command. Where x.x.x.x is ipaddress of the system
- **Chargen :** Generates a stream of ASCII data. Use the **telnet x.x.x.x chargen** command.
- **Discard :** Throws away whatever you type. Use the **telnet x.x.x.x discard** command.
- **Daytime :** Returns system date and time, if it is correct. It is correct if you run Network Time Protocol (NTP), or have set the date and time manually from the exec level. Use the **telnet x.x.x.x daytime** command.

UDP Small Servers

The UDP small servers are:

- **Echo :** Echoes the payload of the datagram you send.
- **Discard :** Silently pitches the datagram you send.
- **Chargen :** Pitches the datagram you send, and responds with a 72-character string of ASCII characters terminated with a CR+LF.

TIPS

The commands to enable for the TCP and UDP small servers are service tcp-small-servers and service udp-small-servers. They may be disabled using the commands no service tcp-small-servers and no service udp-small-servers.

5.3. BYTE ORDERING

Port numbers and IP Addresses are represented by multi-byte data types which are placed in packets for the purpose of routing and multiplexing. Port numbers are two bytes (16 bits) and IP4 addresses are 4 bytes (32 bits), now the problem arises when transferring multi-byte data types between different architectures.

Say Host A uses a "big-endian" architecture and sends a packet across the network to Host B which uses a "little-endian" architecture. If Host B looks at the address to see if the packet is for him/her (choose a gender!), it will interpret the bytes in the opposite order and will wrongly conclude that it is not his/her packet.

The Internet uses big-endian and we call it the network-byte-order,and it is really not important to know which method it uses since we have the following functions to convert host-byte-ordered values into network-byte-ordered values and vice versa:

Example : To convert port numbers (16 bits) from host to network and vice a versa:

```
Host -> Network
unit16_t htons( uint16_t hostportnumber )
Network -> Host
unit16_t ntohs( uint16_t netportnumber )
```

To convert IP4 Addresses (32 bits) from host to network and vice a versa:

```
Host -> Network
unit32_t htonl( uint32_t hostportnumber )
Network -> Host
Unit32_t ntohl( uint32_t netportnumber )
```

Address Structures, Ports, Address conversion functions

5.4. OVERVIEW OF IP4 ADDRESSES

IP4 addresses are 32 bits long. They are expressed commonly in what is known as dotted decimal notation. Each of the four bytes which makes up the 32 address are expressed as an integer value. (0 – 255) and separated by a dot. For example, 138.23.44.2 is an example of an IP4 address in dotteddecimal notation. There are conversion functions which convert a 32 bit address into a dotted decimal string and vice versa.

Often times though the IP address is represented by a domain name, for example, hill.ucr.edu. The importance of IP addresses follows from the fact that each host on the Internet has a unique IP address. Thus, although the Internet is made up of many networks of networks with many different types of architectures and transport mediums, it is the IP address which provides a cohesive structure so that at least the, any two hosts on the Internet can communicate with each other.

HIGHLIGHTS

1. Echo protocol is used for measurement of round trip time (RIT).
2. Following are TCP small servers :
   - Echo
   - Chargen
   - Discard
   - Daytime
3. Following are UDP small servers
   - Echo
   - Discard
   - Chargent
4. Port numbers are 16 bits and IP addresses are 32 bits.

5. With the help of byte ordering conversion from port address to IP address or vice a versa is possible.

REVIEW QUESTIONS

1. What is byte ordering function?
2. What is echo service ?
3. What are TCP small server ?
4. What are UDP small server ?
5. Explain following :
   (i) Discard
   (ii) Chargen

❑❑❑

UNIT-III

6. ALGORITHMS AND ISSUES IN SERVER SOFTWARE DESIGN

Algorithms and Issues in Server Software Design

Chapter 6

INSIDE THIS CHAPTER

6.1. INTRODUCTION [MDU Dec. 2008; Dec. 2006]

A computer or software package that provides a specific kind of service to client software running on other computers is called server.

Servers can be classified based on the basis of following criteria

1. Their servicing discipline (***iterative*** or ***concurrent***);
2. Their communication methods (***connection-oriented*** or ***connectionless***);
3. The information that they keep (***stateless*** or ***stateful***).

There are many ways that a client/server can be designed. Each different algorithm has various benefits and problems. The main client/server algorithms can be classified as:

- iterative connectionless server
- iterative connection oriented server
- concurrent connectionless server
- concurrent connection-oriented server

6.2. BASIC SERVER ALGORITHM

Conceptually each server follows a simple algorithm:

```
it creates a socket
```

```
binds the socket to a well known port
loop
accept the next client
request from this port
serve this request
formulate a reply
send the reply to client
end
```

6.3. PROBLEMS WITH SIMPLE SERVER

1. This is only good enough for simple applications
2. Consider a service requiring considerable time to handle
   - ➢ Each request example suppose an ftp client server were implemented like this
   - ➢ One user requests a huge file
   - ➢ Moments later another user might wish to transfer a small file
3. the second user might has to wait a considerable time **just** to transfer a small file
4. the second user **blocks** until the first user has finished with the server.
5. thus servers are seldom built like this

6.4. ITERATIVE SERVER [MDU Dec. 2008; Dec. 2007]

An *iterative server* serves the requests one after the other.

OR

It is used to describe a server implementation that processes one request at a time

Iterative server are easier to design and implement. Iterative design is suitable when the service time for each request is small (because the mean response time is still acceptably small). It is suitable for simple services such as the TIME service.

Iterative design is not suitable when the service time for a request may be large.

Example :

Two clients are using a file transfer service:

- ➢ the 1st client requests to get a file of size 200 Mbytes,
- ➢ the 2nd client requests to get a file of size 20 bytes.

If iterative design is used, the 2nd client has to wait a long time.

Due to this iterative structure, the processing of each request is serialized at a coarse-grained level, using typically the UNIX **select()** and Windows **WaitForMultipleObjects()** synchronous event demultiplexer. This can under-utilize certain processing resources (multiply CPU or parallel DMA transfer from/to I/O devices).

It could be an other problem, that the client is blocked waiting for the server to process its request.

6.4.1. Pseudo Code for Iterative Server

```
iterative_server()
{
    initialize listener endpoint(s)

    foreach ( new client request )
    {
        retrieve next request from an input source
        perform requested service
        if ( response required )
            send response to client
    }
}
```

6.5. CONCURRENT SERVER [MDU Dec. 2008; May 2008; Dec. 2007]

A concurrent server serves multiple clients simultaneously

OR

It is used to describe a server that handles multiple requests at one time. concurrent servers are more difficult to design and build because

➢ The resulting code is more complex

➢ Difficult to modify the code

Concurrent servers are well-suited for I/O-bound services and/or long- duration services that require variable amounts of time to execute. This design requires synchronization mechanisms such as ***semaphores*** and ***mutex locks***. This is the pseudo-code of a multi-threaded concurrent server:

Concurrency Refers to Either

1. real simultaneous computing (using multiple processors)

   or

2. apparent simultaneous computing (by time sharing)

Concurrent design is more difficult to implement, but it can give a smaller response time when

➢ different clients require very different service time, or

➢ the service requires significant I/O, or

➢ the server is executed on a computer with multiple processors.

However most programmers choose concurrent server implementations as iterative servers

- cause unnecessary delays in distributed applications.
- may be a performance bottleneck that effects many client applications.

TIPS

The term concurrent server refers to whether the server handles multiple requests concurrently, not to whether the underlying implementation uses multiple concurrent processes.

6.5.1. Pseudo Code for Multithreaded Concurrent Server

```
master_thread()
{
    initialize listener endpoint(s)

    foreach ( new client request )
    {
        receive the request
        spawn new worker thread and
pass request
    }
}
worker_thread()
{
    perform requested service
    if ( response required )
        send response to client
    terminate thread
}
```

DO YOU KNOW

Reactive servers process multiply requests virtually simultaneously, although all processing is implemented by a single thread.

6.6. COMPARISON

Iterative server implementations, which are easier to build and understand, may result in poor performance because they make clients wait for service. Whereas in contrast, concurrent server implementations, which are more difficult to build, yield better performance.

DO YOU KNOW

In a single theraded environment there are another way to get back the control. With the help of *asynchronous I/O* the server initiate an input/output request, and starts working on another client request. At the complition of the I/O the operating system notifies the server, which can turn back to the original client and finish the service.

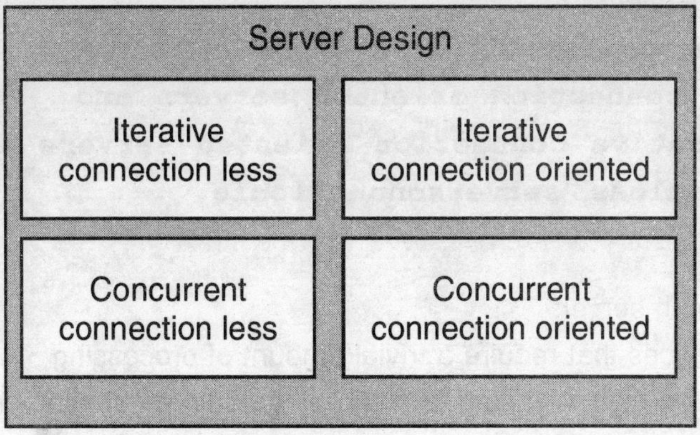

Fig. 6.1.

Iterative Connection Oriented and Connectionless Server

1. TCP provides a *connection-oriented*
2. UDP provides a *connectionless* service
3. Servers that use TCP are, by definition, connection oriented
4. Those that use UDP are Connectionless serversrative connection oriented servers and Iter

6.7. CONNECTIONLESS SERVERS

```
create a socket and bind
to the well known address
of the service being offered.
put socket into passive mode, making it
ready for use by a server
loop
   accept the next connection
   request from the socket,
   and obtain a new socket
   for the connection
repeat
   read a request from
   the client,
   generate a reply,
   send the reply
   aback to the client

close connection
```

until finished with the client ;

```
Iterclose connection
endative connection oriented servers and
cIterIterative connection oriented servers and
connectionless serversonnectionle
```

Drawback

- a less common server
- used for services that require a trivial amount of processing
- it might incur a high overhead in establishing and terminating connections

Following functions are used

Getportbyname

- Use this function to map a particular service onto a well known port number
- Need to ensure that our application port number does not conflict with someone else application
- In UNIX all port numbers are defined in /etc/services

bind

- Use this function to bind a socket to a port number and the servers IP address.

listen

- The server uses this function to place a socket into passive mode
- It also takes an argument which specifies the length of request queue for this socket

accept

- Use this function to obtain the next incoming connection request
- Returns the new descriptor of a socket which can then be used by read and write

close

- To close the connection with the client the server uses close.

6.8. ITERATIVE CONNECTIONLESS SERVER [MDU May 2008; Dec. 2007; May 2006]

Iterative servers work best for services that have a low request processing time. the connection oriented protocol TCP has a high connection and disconnection overhead thus most iterative servers use connectionless protocol such as UDP

Some iterative servers use UDP because

(1) Iterative design is suitable for services with a small service time,

(2) UDP has a lower overhead than TCP.

- The server uses an *unconnected socket* to receive datagrams from arbitrary clients.

➢ The server calls *recvfrom()* to receive a client's request and extracts the client's endpoint address.

➢ The server calls *sendto()* to send a reply message to the client's endpoint address.

```
create a socket and bind
to a well known address
for which a service is
being offered
loop
read next request from client
process the request
send reply back to client
end
```

Advantages

1. The most common form of connectionless server, used especially for service that require a trivial amount of processing for each request
2. Iterative servers are often stateless, making them easier to understand and less vulnerable to failures

Following functions are used :

Sendto-use this function to send data across a connectionless socket.

Recvfrom use this function to receive data from a connectionless socket.

Concurrent connection oriented server and connection less server.

Primary reason for introducing concurrency is that provide faster response time to multiple clients works well when forming a response requires significant input output the processing time required varies dramatically

```
create a socket and bind
to the well known address
for the service being offered
leave the socket unconnected
loop
call recvfrom to obtain the
next client request
if (fork() == 0) {
/* child process */
do whatever the
client request says
form a reply and send
```

```
it to client
(use sendto)
exit
} else {
/* must be the parent */
}
End
```

The most general type of server because it offers reliable transport as well as the ability to handle multiple requests at the same time.

TIPS

Connection oriented servers implement concurrency among connections rather than individual requests.

```
create a socket and bind
   it to the well known address
   for the service being offered
   place socket into passive mode
   making it ready for use by
   the server
loop
   call accept to receive the
   next request from a client
   if (fork() == 0) {
   /* must be the child */
repeat
   read request from client
   do whatever the client
   request says
   form a reply and send
   it to client
   until client wishes to quit
   close connection
exit
   } else {
   /* must be the parent */
   }
End
```

6.9. WHEN TO USE EACH SERVER TYPE [MDU May 2008; May 2006]

6.9.1. Iterative vs Concurrent

Iterative easier to design, implement and maintain but concurrent can provide quicker response to requests.use iterative implementation if request processing time is short and the code produces fast responses times

6.9.2. Connection Oriented vs Connectionless

Connection oriented access means using TCP.It implies reliable delivery because connectionless transport means using UDP it implies unreliable delivery.only use connectionless transport if the application protocol handles reliability or the local area network exhibits:

- ➢ Low packet loss
- ➢ No packet reordering (very few do)

Use connection oriented transport whenever a wide area network separates client and server. Never move a connectionless client and server to a wide area network without checking to see if the application protocol handles the reliability problems.

Summary

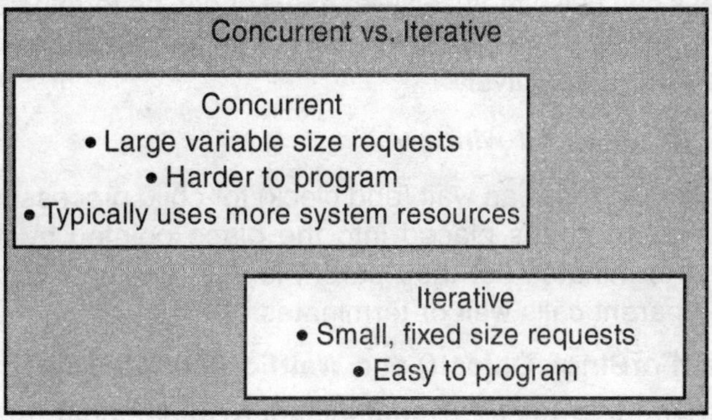

Fig. 6.2.

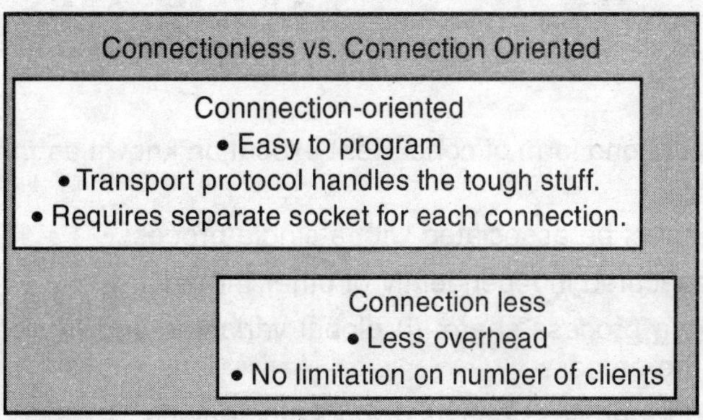

Fig. 6.3.

6.10. PROCESS

We can use *processes* and *threads* to realize concurrency on a processor via time sharing.

A process is a fundamental unit of computation. The OS has to store various information about each process.

Operation on process in unix and window.

1. Creating Process in Unix and Window

On UNIX systems new process is created using the **fork()** call. Fork copies all data the process owned except the **pid**. That means all open descriptors (files, semaphores, sockets, etc..) are inherited from the parent process as well as the full address space with all the variables.

Once a process forked - and received a new pid - it may mutating to execute a different binary with the system calls from the family of **exec*** functions. Here kernel space of the process remain intact: therefore we can use the handlers, descriptors, etc. from the process called the exec.

On Windows we call **CreateProcess()** with vrious parameters to create a new process.

2. Terminating process in unix and window

On UNIX a process can finish themself with **exit()** or can be terminated - with the required authorization - by the **kill()** system call. In Windows that can be done with **ExitProcess()** or with **TerminateProcess()** respectively.

3. Waiting process in unix and window

On UNIX systems a process can wait (and block) for child processes with the **wait()** and **waitpid()** calls. The return status placed into the place pointed by the argument of the function. If a process terminates but their parent is still running it becames a *zombie*, its state is kept until the parent calls wait or terminates.

On Windows **WaitForSingleObject()** and **WaitForMultipleObject()** has a similar role.

Many other IPC (Inter Process Communication) techniques are exist to establish communication between processis, including **semaphores**, **signals**, **messages** and (internal) **sockets**.

6.10.1. Thread

Windows provides a second form of concurrent execution known as *threads of execution* or *threads.*

- ➢ Each thread must be associated with a single process.
- ➢ A thread is executed independently of other threads.
- ➢ All threads in a process share: (i) global variables and (ii) resources that the OS allocates to the process.
- ➢ Each thread in a process has its own local variables.

For example, if multiple threads execute the following piece of code :

```
for ( i=1 ; i<=10 ; i++ )
printf ( "%d\n" , i ) ;
then each thread has its own index variable i.
```

6.10.2. Difference between Process and Thread

Process is an OS entity that provides the context for executing program instructions and managing resources, like *memory*, *I/O handles*, *signals* etc. A process is **protected** from other OS processes via memory management (today supported by hardware). Every process has its own *address space*.

However starting a new process is a highly expensive procedure in most operating systems. One can use **fork()** system call on UNIX systems, which copies the whole process (data). On Windows **CreateProcess()** could be used. Communication, data exchange between processes require special tools, **Inter Process Communication**, *IPC* and could be also expensive.

Starting a thread in the same time is much cheaper with **pthread_create()** in UNIX systems and **CreateThread()** on Win32. Threads share the same OS address space, so data exchange is cheap.

However threads are not "protected": blocking or aborting one thread could be affecting the whole program. It is not necessary that multithreading always increases performance p. Many times performance doesn't improved for several reasons, like:

- Compute-bound applications on a uniprocessor won't benefit from multithreading, since computation and communication can't run in parallel.
- Fine-grained locking strategies can yield high synchronization overhead

6.11. IMPLEMENTING CONCURRENT SERVER

A concurrent server executes a ***main thread*** and multiple ***service threads***.

Main Thread (or Master Thread)

- When a server program starts execution, the main thread is executed.
- The main thread creates a socket and waits for clients' requests.
- When a client's request arrives, the main thread creates a new service thread to serve this client. Meanwhile, the main thread waits for another request.

Service Thread (or Slave Thread)

- Each service thread is created to serve one client.
- When the service is completed, the service thread is terminated.
- If the server computer is serving *N* clients, there are one main thread and *N* service threads.

Connection-Oriented Servers

A *connection-oriented server* uses TCP for connection-oriented communication.

TCP provides reliable transport.

➢ The servers need not do so.

➢ The servers are simpler.

A server uses a separate socket for each connection.

Connection-less Servers

A *connectionless server* uses UDP for connectionless communication.

UDP does not guarantee reliable transport.

➢ A connectionless server may need to realize reliability through timeout and retransmission.

➢ The server and client are relatively complicated.

6.12. STATELESS SERVER

A *stateless server* does not keep information about the ongoing interactions with each client.

Example :

➢ Consider the service that a client can read a file stored on the server's computer.

➢ To maintain stateless, each client's request must specify a file name, a position in the file, and the number of bytes to read.

➢ The server handles each request independently:

- open the specified file,
- seek to the specified position,
- read the specified number of bytes,
- send these bytes to the client,
- close the file.

TIPS

Stateless servers are less efficient but they are more reliable they may be a good choice for UDP transport.

6.13. STATEFUL SERVERS

A *stateful server* keeps information about the status of ongoing interactions with each client.

Examples

Consider the file-reading service.

➢ When a client crashes and it is restarted, it is likely that it will adopt another protocol port number the table entry and buffer allocated earlier are wasted.

➢ The server may receive the client's requests in a different order (e.g., using UDP) the server returns the bytes of the file in a wrong order.

TIPS

When clients / computers / networks fail, the server performance may be degraded.

6.14. MULTIPROTOCOL SERVER

[MDU Dec. 2007; Dec. 2006]

A given server handles requests for one particular service through one particular protocol Example: DAYTIME service runs two servers – one for UDP request another for TCP requests. Main disadvantage of using one server per protocol is replication.

A multiprotocol server consists of a single process that uses asynchronous I/O to handle UDP/TCP.

Multiprotocol Server opens two sockets (UDP/TCP). It Uses asynchronous I/O to wait for one of the sockets to become ready.

- TCP socket ready – client requested a TCP connection
- UDP socket ready – client requested a UDP connection

TIPS

A multiprotocol server design permits the designer to create a single procedure that responds to requests for a given service and to call that procedure regardless of whether requests arrive via UDP or TCP.

Process Structure

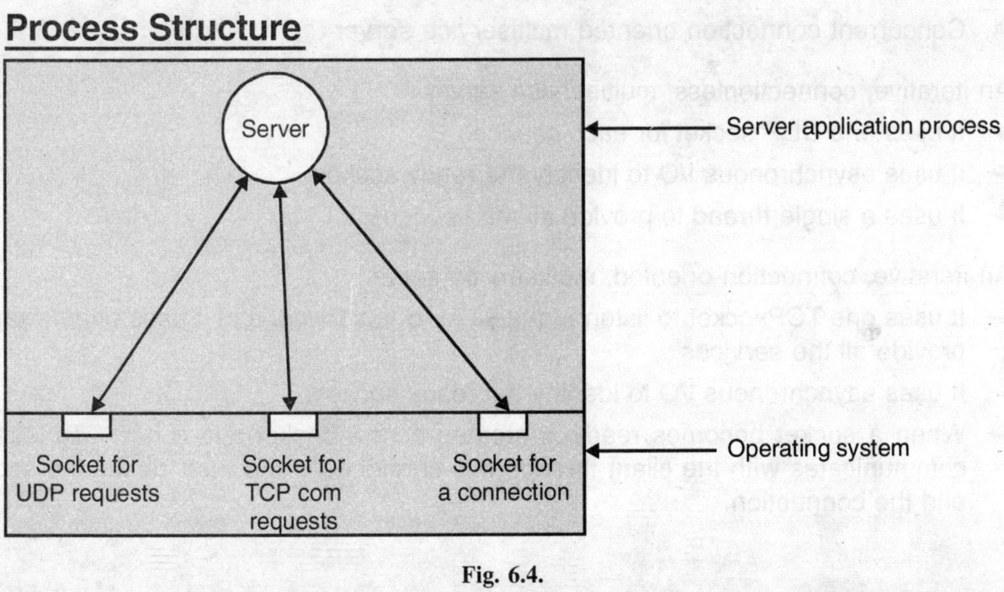

Fig. 6.4.

A service can use TCP, or UDP, or both.If we implement one server program using TCP and another server program using UDP, then

(1) It is more difficult to manage and maintain two programs,

(2) They require more resources during execution.

It is desirable to implement one server program that can provide different transport for different clients' requests this server is called a *multiprotocol server.*

There are many design choices for multiprotocol servers

1. iterative design,
2. concurrent design.

An iterative multiprotocol server -The server creates a TCP socket and a UDP socket.

➢ The server uses asynchronous I/O to identify the ready socket.

➢ If the TCP socket becomes ready, the server executes *accept()* to create a new TCP connection and then communicates with the client over this connection.

➢ If the UDP socket becomes ready, the server calls *recvfrom()* to receive the message and the sender's endpoint address, and then calls *sendto()* to send the response back to the client.

6.15. MULTISERVICE SERVER [MDU May 2008; Dec. 2007; Dec. 2006; May 2006]

A *multiservice server* provides multiple services, so that it is only necessary to maintain one server program.

There are many design choices for multiservice server

1. Iterative connectionless multiservice server
2. Iterative connection oriented multiservice server
3. Concurrent connectionless multiservice server
4. Concurrent connection oriented multiservice server

An iterative, connectionless, multiservice server :

➢ It uses one UDP socket for each service.

➢ It uses asynchronous I/O to identify the ready sockets.

➢ It uses a single thread to provide all the services.

An iterative, connection-oriented, multiservice server

➢ It uses one TCPsocket to listen requests for one service, and it uses one thread to provide all the services.

➢ It uses asynchronous I/O to identify the ready sockets.

➢ When a socket becomes ready, it creates a new socket and a new connection, communicates with the client through this connection, and then closes the socket and the connection.

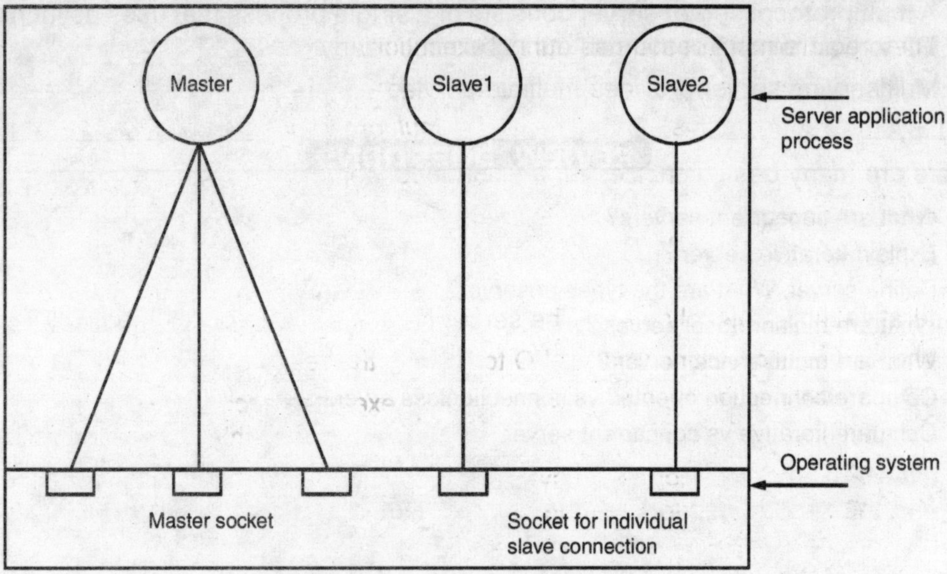

Fig. 6.5. Multiservice Server.

A Concurrent, Connection-Oriented, Multiservice Server

- The master thread creates one TCP socket to listen requests for one service, and it uses asynchronous I/O to identify the ready sockets.
- When the master thread receives a client's request, it creates a slave thread to serve this client.

HIGHLIGHTS

1. A software or package that provides specific kind of service to client software running on either computer is called server.
2. Servers can be classified as :
   - Iterative connectionless server.
   - Iterative connection oriented server.
   - Concurrent connectionless server.
   - Concurrent connection-oriented services.
3. An iterative server serves the requests one after the other.
4. A concurrent server servers multiple clients simultaneously.
5. Iterative are easier to design, implement but concurrent can provide quicker response to requests.
6. Connection-oriented means using TCP (reliable).
7. Connection-less means using UDP (unreliable)
8. A stateful server keeps information about the status of ongoing interactions with each client.

9. A multiprotocol server server consists of a single process that uses asynchronous I/O to wait for one of the sockets to become ready.
10. Multiservice server provides multiple service.

REVIEW QUESTIONS

1. What are concurrent servers?
2. Explain iterative server.
3. Define server. What are the types of server?
4. What are multiprotocol server ?
5. What are multiservice server ?
6. Compare connection oriented vs connectionless erver.
7. Compare iterative vs concurrent server.

❑❑❑

UNIT-IV

7. REMOTE PROCEDURE CALLING

8. RPC AUTHENTICATION

Remote Procedure Call Concept

Chapter 7

INSIDE THIS CHAPTER

7.1. INTRODUCTION

[MDU Dec. 2008, May 2008; May 2006]

RPC is a powerful technique for constructing distributed, client-server based applications. It is based on extending the notion of conventional, or local procedure calling, so that the called procedure need not exist in the same address space as the calling procedure. The two processes may be on the same system, or they may be on different systems with a network connecting them. By using RPC, programmers of distributed applications avoid the details of the interface with the network. The transport independence of RPC isolates the application from the physical and logical elements of the data communications mechanism and allows the application to use a variety of transports.

> **DO YOU KNOW**
>
> **RPC makes the client/server model of computing more powerful and easier to program.**

> **DO YOU KNOW**
>
> **RPC uses the client/server model. The requesting program is a client and the service-providing program is the server.**

Like a regular or local procedure call, an RPC is a synchronous operation requiring the requesting program to be suspended until the results of the remote procedure are returned. However, the use of *lightweight processes* or threads that share the same address space allows multiple RPCs to be performed concurrently.

TIPS

Remote Procedure Call (RPC) is a protocol that one program can use to request a service from a program located in another computer in a network without having to understand network details.

7.2. WORKING OF RPC

[MDU Dec. 2008; May 2008]

An RPC is analogous to a function call. Like a function call, when an RPC is made, the calling arguments are passed to the remote procedure and the caller waits for a response to be returned from the remote procedure. Folowing Figure shows the flow of activity that takes place during an RPC call between two networked systems. The client makes a procedure call that sends a request to the server and waits. The thread is blocked from processing until either a reply is received, or it times out. When the request arrives, the server calls a dispatch routine that performs the requested service, and sends the reply to the client. After the RPC call is completed, the client program continues. RPC specifically supports network applications.

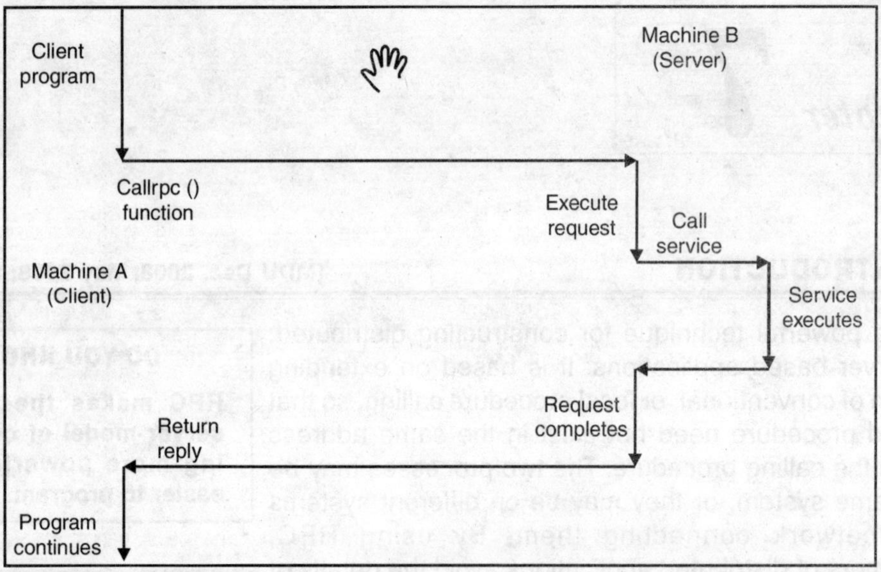

Fig. 7.1.

7.3. REMOTE PROCEDURE CALLING MECHANISM

A remote procedure is uniquely identified by the three unsigned integer field:

➢ program number,

➢ version number,

➢ procedure number

DO YOU KNOW

A procedure call is also sometimes known as a function call or a subroutine call.

The program number identifies a group of related remote procedures, each of which has a unique procedure number. Program numbers are administered by some central authority (like Sun). Once an implementor has a program number, he can implement his remote program; the first implementation would most likely have the version number of 1. Because most new protocols evolve into better, stable, and mature protocols.

A program may consist of one or more versions. Each version consists of a collection of procedures which are available to be called remotely. Version numbers enable multiple versions of an RPC protocol to be available simultaneously. Each version contains a a number of procedures that can be called remotely.

TIPS

Version Numbers typically start at 1 and are numbered sequentially.

Each procedure has a procedure number. The procedure number identifies the procedure to be called. These numbers are documented in the specific program's protocol specification. For example, a file service's protocol specification may state that its procedure number 5 is "read" and procedure number 12 is "write".

TIPS

Procedure Identifiers usually start at 1 and are numbered sequentially

Just as remote program protocols may change over several versions, the actual RPC message protocol could also change. Therefore, the call message also has in it the RPC version number, which is always equal to two for the version of RPC described here.

The reply message to a request message has enough information to distinguish the following error conditions:

The remote implementation of RPC does speak protocol version 2.

(1) The lowest and highest supported RPC version numbers are returned.

(2) The remote program is not available on the remote system.

(3) The remote program does not support the requested version number. The lowest and highest supported remote program version numbers are returned.

(4) The requested procedure number does not exist. (This is usually a caller side protocol or programming error.)

(5) The parameters to the remote procedure appear to be garbage from the server's point of view.

7.4. RPC MODEL [MDU Dec. 2008; Dec. 2007; Dec. 2006]

A client application issues a normal procedure call to a *client stub.* The client stub receives arguments from the calling procedure and returns arguments to the calling procedure. An argument may instantiate an input parameter, an output parameter, or an input/output parameter. In the discussion of this Section, the term *input argument* refers to a parameter which may be either an input parameter or an input/output parameter, and the term *output argument* refers to either an output parameter or an input/output parameter.

The client stub converts the input arguments from the local data representation to a common data representation, creates a message containing the input arguments in their common data representation, and calls the client runtime, usually an object library of routines that supports the functioning of the client stub. The client runtime transmits the message with the input arguments to the server runtime which is usually an object library that supports the functioning of the server stub. The server runtime issues a call to the *server stub* which takes the input arguments from the message, converts them from the common data representation to the local data representation of the server, and calls the server application which does the processing

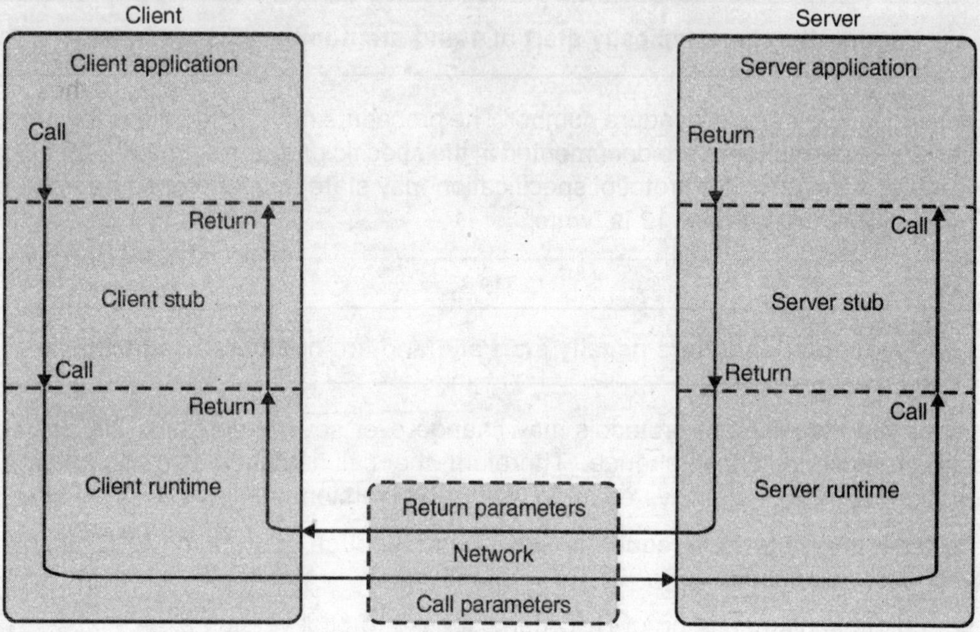

Fig. 7.2.

When the server application has completed, it returns to the server stub the results of the processing in the output arguments. The server stub converts the output arguments from the data representation of the server to the common data representation for transmission on the network and encapsulates the output arguments into a message which is passed to the server runtime. The server runtime transmits the message to the client runtime which passes the message to the client stub. Finally, the client stub extracts the arguments from the message and returns them to the calling procedure in the required local data representation.

Like a normal procedure call, RPC is a synchronous operation, i.e., the client process is blocked until processing by the server is complete. This is not acceptable for many applications. As a consequence, the RPC model is enhanced to include the concept of a *lightweight process*. A lightweight process (also known as a *thread*) is an independent execution path within a normal process. A normal process can consist of several lightweight processes, each behaving like a normal process from the point of view of CPU use. However, all lightweight processes of the same process share the same address space. Thus, context switches between lightweight processes may be done more economically than context switches between normal processes.

In order to achieve asynchronous operation, a client application initiates an RPC call in a lightweight process and then proceeds with other processing. The application can recognize the completion of the RPC by some technique such as a status check or a software interrupt

Sequence of Events During a RPC

1. The client calls the Client stub. The call is a local procedure call, with parameters pushed on to the stack in the normal way.
2. The client stub packs the parameters into a message and makes a system call to send the message. Packing the parameters is called marshalling.
3. The kernel sends the message from the client machine to the server machine.
4. The kernel passes the incoming packets to the server stub.
5. Finally, the server stub calls the server procedure. The reply traces the same in other direction.

7.5. RPC-SUPPORTED NETWORK PROTOCOLS

The following table lists the network protocols supported by RPC and the type of RPC connection for which the protocol is used.

Protocol	RPC Type
Transmission Control Protocol (TCP)	Connection–oriented
Sequenced Packet Exchange (SPX)	Connection–oriented
Named Pipe	Connection–oriented
HTTP	Connection–oriented
User Datagram Protocol (UDP)	Connectionless
Cluster Datagram Protocol (CDP)	Connectionless

7.5.1. Network Ports Used by RPC

RPC server programs typically use dynamic port mappings to avoid conflicts with programs and protocols registered in the range of well-known TCP ports. RPC server programs associate their universally unique identifier (UUID) with a dynamic port and register the combination with the RPC EPM. The EPM provides a single point of contact for RPC clients. The RPC clients contact the EPM and use the server program's UUID to determine the port being used by the server program. The following table indicates the network ports normally used by RPC.

7.5.2. Port Assignments for RPC

Service Name	UDP	TCP
HTTP	80, 443, 593	80, 443, 593
Named Pipes	445	445
RPC Endpoint Mapper	135	135
RPC Server Programs	<Dynamically assigned>	<Dynamically assigne

Syntax Notes

1. The following keywords are added and cannot be used as identifiers: “program” and “version”;
2. A version name cannot occur more than once within the scope of a program definition. Nor can a version number occur more than once within the scope of a program definition.
3. A procedure name cannot occur more than once within the scope of a version definition. Nor can a procedure number occur more than once within the scope of version definition.
4. Program identifiers are in the same name space as constant and type identifiers.
5. Only unsigned constants can be assigned to programs, versions and procedures.

7.6. RPC DYNAMIC PORT ALLOCATION MAPPING

[MDU Dec. 2008; May 2008; Dec. 2007; May 2006]

To solve the port identification problem, a client must be able to map from an RPC program and a machine address to the protocol port that the server obtained on the destination machine when it started. The mapping must be dynamic because it can change if the machine reboots or if the RPC program starts execution again.

To allow clients to contact remote programs, the Sun RPC mechanism includes a dynamic mapping service. The RPC port mapping mechanism uses a server to maintain a small database of port mappings on each machine. This RPC server waits on a particular port number (111) and it receives the requests for all remote procedure calls.

Whenever a remote program (i.e., a server) begins execution, it allocates a local port that it will use for communication. The remote program then contacts the server on its local machine for registration and adds a pair of integers to the database:

```
(RPC program number, protocol port number)
```

Once an RPC program has registered itself, callers on other machines can find its protocol port by sending a request to the server. To contact a remote program, a caller must know the address of the machine on which the remote program executes as well as the RPC program number assigned to the program. The caller first contacts the server on the target machine, and sends an RPC program number. The server returns the protocol port number that the specified program is currently using.

This server is called the ***RPC port mapper*** or simply the ***port mapper***. A caller can always reach the port mapper because it communicates using the well known protocol port, 111. Once a caller knows the protocol port number the target program is using, it can contact the remote program program directly.

TIPS

RPC dynamic port allocation is used by DHCP manager.

Remote Procedure Call (RPC) dynamic port allocation is used by remote administration applications such as Dynamic Host Configuration Protocol (DHCP) Manager, Windows Internet Name Service (WINS) Manager, and so on. RPC dynamic port allocation will instruct the RPC program to use a particular random port above 1024.

Customers using firewalls may want to control which ports RPC is using so that their firewall router can be configured to forward only these Transmission Control Protocol (TCP) ports.

7.7. AUTHENTICATION AND AUTHORIZATION [MDU May 2008]

In applications, authentication is the process that confirms the identity of the user. Usually this takes the form of a login form, HTTP basic authentication, or something similar. Authorization is the process to determine whether the user has sufficient privileges to perform the specified action. This takes the form of permission checks based on the authenticated user. Therefore, authentication must come before authorization.

TIPS

Authentication is the process that confirms the identity of user.

Provisions for authentication of caller to service and vice-versa are provided as a part of the RPC protocol. The call message has two authentication fields,

- the credentials
- and verifier.

The reply message has one authentication field, the response verifier. The RPC protocol specification defines all three fields to be the following :

```
opaque type:

        enum auth_flavor {
            AUTH_NULL       = 0,
            AUTH_UNIX       = 1,
            AUTH_SHORT      = 2,
            AUTH_DES        = 3
            /* and more to be defined */
        };

        struct opaque_auth {
            auth_flavor flavor;
            opaque body<400>;
        };
```

In simple English, any "opaque_auth" structure is an "auth_flavor" enumeration followed by bytes which are opaque to the RPC protocol implementation. The interpretation and semantics of the data contained within the authentication fields is specified by individual, independent authentication protocol specifications. If authentication parameters were rejected, the response message contains information stating why they were rejected.

7.7.1. Program Number Assignment

Program numbers are given out in groups of hexadecimal 20000000 (decimal 536870912) according to the following chart:

```
       0 - 1fffffff   defined by Sun
20000000 - 3fffffff   defined by user
40000000 - 5fffffff   transient
60000000 - 7fffffff   reserved
80000000 - 9fffffff   reserved
a0000000 - bfffffff   reserved
c0000000 - dfffffff   reserved
e0000000 - ffffffff   reserved
```

The first group is a range of numbers administered by Sun Microsystems and should be identical for all sites. The second range is for applications peculiar to a particular site. This range is intended primarily for debugging new programs. When a site develops an application that might be of general interest, that application should be given an assigned number in the first range. The third group is for applications that generate program numbers dynamically. The final groups are reserved for future use, and should not be used.

Other Uses of the RPC Protocol

The intended use of this protocol is for calling remote procedures. That is, each call message is matched with a response message.However, the protocol itself is a message-passing protocol with which other (non-RPC) protocols can be implemented. Sun currently uses, or perhaps abuses, the RPC message protocol for the following two (non-RPC) protocols:

- batching (or pipelining)
- broadcast RPC.

Batching

Batching allows a client to send an arbitrarily large sequence of call messages to a server; batching typically uses reliable byte stream protocols (like TCP/IP) for its transport. In the case of batching, the client never waits for a reply from the server, and the server does not send replies to batch requests. A sequence of batch calls is usually terminated by a legitimate RPC in order to flush the pipeline (with positive acknowledgement).

Broadcast RPC

In broadcast RPC-based protocols, the client sends a broadcast packet to the network and waits for numerous replies. Broadcast RPC uses unreliable, packet-based protocols (like UDP/IP) as its transports. Servers that support broadcast protocols only respond when the request is successfully processed, and are silent in the face of errors. Broadcast RPC uses the Port Mapper RPC service to achieve its semantics.

7.8. AUTHENTICATION PROTOCOLS

Authentication parameters are opaque, but open-ended to the rest of the RPC protocol.

7.8.1. Null Authentication

Often calls must be made where the caller does not know who he is or the server does not care who the caller is. In this case, the flavor value (the discriminant of the opaque_auth's union) of the RPC message's credentials, verifier, and response verifier is "AUTH_NULL". The bytes of the opaque_auth's body are undefined. It is recommended that the opaque length be zero.

7.8.2. UNIX Authentication

The caller of a remote procedure may wish to identify himself as he is identified on a UNIX(tm) system. The value of the credential's discriminant of an RPC call message is "AUTH_UNIX". The bytes of the credential's opaque body encode the the following structure:

```
struct auth_unix {
    unsigned int stamp;
    string machinename<255>;
    unsigned int uid;
    unsigned int gid;
    unsigned int gids<10>;
};
```

The "stamp" is an arbitrary ID which the caller machine may generate.The "machinename" is the name of the caller's machine (like "krypton"). The "uid" is the caller's effective user ID. The "gid" is the caller's effective group ID. The "gids" is a counted array of groups which contain the caller as a member. The verifier accompanying the credentials should be of "AUTH_NULL" (defined above).

The value of the discriminant of the response verifier received in the reply message from the server may be "AUTH_NULL" or "AUTH_SHORT".

In the case of "AUTH_SHORT", the bytes of the response verifier's string encode an opaque structure. This new opaque structure may now be passed to the server instead of the original "AUTH_UNIX" flavor credentials. The server keeps a cache which maps shorthand opaque structures (passed back by way of an "AUTH_SHORT" style response verifier) to the original credentials of the caller. The caller can save network bandwidth and server cpu cycles by using the new credentials.

The server may flush the shorthand opaque structure at any time. If this happens, the remote procedure call message will be rejected due to an authentication error. The reason for the failure will be "AUTH_REJECTEDCRED". At this point, the caller may wish to try the original "AUTH_UNIX" style of credentials.

7.8.3. DES Authentication

UNIX authentication suffers from two major problems:

(1) The naming is too UNIX oriented.

(2) There is no verifier, so credentials can easily be faked.

DES authentication attempts to fix these two problems.

1. Naming

The first problem is handled by addressing the caller by a simple string of characters instead of by an operating system specific integer. This string of characters is known as the "netname" or network name of the caller. The server is not allowed to interpret the contents of the caller's name in any other way except to identify the caller. Thus, netnames should be unique for every caller in the Internet.

It is up to each operating system's implementation of DES authentication to generate netnames for its users that insure this uniqueness when they call upon remote servers. Operating systems already know how to distinguish users local to their systems. It is usually a simple matter to extend this mechanism to the network.

For example, a UNIX user at Sun with a user ID of 515 might be assigned the following netname: "unix.515@sun.com".

This netname contains three items that serve to insure it is unique. Going backwards, there is only one naming domain called "sun.com" in the Internet. Within this domain, there is only one UNIX user with user ID 515. However, there may be another user on another operating system, for example VMS, within the same naming domain that, by coincidence, happens to have the same user ID. To insure that these two users can be distinguished, we add the operating system name. So, one user is "unix.515@sun.com" and the other is "vms.515@sun.com".

The first field is actually a naming method rather than an operating system name. It just happens that today, there is almost a one-to-one correspondence between naming methods and operating systems. If the world could agree on a naming standard, the first field could be the name of that standard, instead of an operating system name.

2. DES Authentication Verifier

Unlike UNIX authentication, DES authentication does have a verifier so the server can validate the client's credential (and vice-versa).

The contents of this verifier is primarily an encrypted timestamp. The server can decrypt this timestamp, and if it is close to what the real time is, then the client must have encrypted it correctly. The only way the client could encrypt it correctly is to know the "conversation key" of the RPC session. And, if the client knows the conversation key, then it must be the real client.

The conversation key is a DES key which the client generates and notifies the server of in its first RPC call. The conversation key is encrypted using a public key scheme in this first transaction.The particular public key scheme used in DES authentication is Diffie-Hellman , with 128-bit keys.

The client and the server need the same notion of the current time in order for all of this to work. If network time synchronization cannot be guaranteed, then client can synchronize with the server before beginning the conversation, perhaps by consulting the Internet Time Server.

The way a server determines if a client timestamp is valid is somewhat complicated. For any other transaction but the first, the server just checks for two things:

(1) the timestamp is greater than the one previously seen from the same client.

(2) the timestamp has not expired.

A timestamp is expired if the server's time is later than the sum of the client's timestamp, plus what is known as the client's "window". The "window" is a number the client passes (encrypted) to the server in its first transaction. You can think of it as a lifetime for the credential.

This explains everything but the first transaction. In the first transaction, the server checks only that the timestamp has not expired. If this was all that was done though, then it would be quite easy for the client to send random data in place of the timestamp with a fairly good chance of succeeding. As an added check, the client sends an encrypted item in the first transaction known as the "window verifier" which must be equal to the window minus 1, or the server will reject the credential.

The client too, must check the verifier returned from the server to be sure it is legitimate. The server sends back to the client the encrypted timestamp it received from the client, minus one second. If the client gets anything different than this, it will reject it.

HIGHLIGHTS

1. RPC is a technique for constructing distributed client-server based application.
2. RPC is analogus to function call.
3. A remote procedure is uniquely identified by three unsigned integer field.
   - Program number
   - Version number
   - procedure number
4. Program number are hexadecimal number
5. RPC supports various protocol
   - TCP
   - SPX

- ➢ HTTP
- ➢ UDP
- ➢ CDP

6. RPC provides dynamic port allocation.
7. Authentication is the process that confirms the identity of user.
8. RPC authentication has two fields :
   - ➢ Credentials
   - ➢ Verifier

REVIEW QUESTIONS

1. Explain the working of RPC.
2. Discuss RPC model.
3. Explain remote procedure calling mechanism.
4. What is dynamic port allocation/mapping ?
5. What is RPC authentication.
6. What are various authentication protocol ?

❑❑❑

RPC Authentication

Chapter 8

INSIDE THIS CHAPTER

8.1. INTRODUCTION [MDU May 2008]

The caller may not want to identify itself to the server, and the server may not require an ID from the caller. However, some network services, such as the Network File System (NFS), require stronger security. Remote Procedure Call (RPC) authentication provides a certain degree of security.

Secure RPC is based on a combination of

➢ Public key cryptography and
➢ Secret key cryptography,

TIPS

Sun's implementation uses the Diffie-Hellman mechanism for key exchange between users, and DES secret key cryptography for encrypting information that is sent over the network.

TIPS

DES is also used to encrypt the user's secret key that is stored in a central network server. This encryption eliminates the need for users to memorize or carry around the hundred-digit numbers that make up their secret keys.

Secure RPC solves many of the problems of AUTH_UNIX style authentication. Because both users and computers must be authenticated, it eliminates many of the spoofing problems to which other systems lend themselves. Indeed, when used with higher-level protocols, such as NFS, Secure RPC can bring unprecedented security to the networked environment

The following are part of RPC authentication:

- RPC Authentication Protocol
- NULL Authentication
- UNIX® Authentication
- Data Encryption Standard (DES) Authentication
- DES Authentication Protocol
- Diffie-Hellman Encryption

DO YOU KNOW

RPC deals only with authentication and not with access control of individual services.

Each service must implement its own access control policy and reflect this policy as return statuses in its protocol. The programmer can build additional security and access controls on top of the message authentication.

The authentication subsystem of the RPC package is open-ended. Different forms of authentication can be associated with RPC clients. That is, multiple types of authentication are easily supported at one time. Examples of authentication types include :

- UNIX,
- DES, and
- NULL.

The default authentication type is none (AUTH_NULL).

8.2. RPC AUTHENTICATION PROTOCOL [MDU May 2008]

The RPC protocol provisions for authentication of the caller to the server, and vice versa, are provided as part of the RPC protocol. Every remote procedure call is authenticated by the RPC package on the server. Similarly, the RPC client package generates and sends authentication parameters. The call message has two authentication fields: credentials and verifier. The reply message has one authentication field: response verifier.

The following RPC protocol specification defines as an opaque data type the credentials of the call message and the verifiers of both the call and reply messages:

```
enum auth_flavor {
    AUTH_NULL     = 0,
    AUTH_UNIX     = 1,
    AUTH_SHORT    = 2,
    AUTH_DES      = 3
    /* and more to be defined */
};
struct opaque_auth {
    auth_flavor flavor;
    opaque body<400>;
};
```

Any **opaque_auth** structure is an **auth_flavor** enumeration followed by bytes that are opaque to the RPC protocol implementation. The interpretation and semantics of the data contained within the authentication fields are specified by individual, independent authentication protocol specifications.

If authentication parameters are rejected, response messages state the reasons. A server can support multiple types of authentication at one time.

8.3. NULL AUTHENTICATION [MDU May 2008]

Sometimes, the RPC caller does not know its own identity or the server does not need to know the caller's identity. In these cases, the **AUTH_NULL** authentication type can be used in both the call message and response messages. The bytes of the opaque_auth body are undefined. The opaque length should be 0.

8.4. UNIX AUTHENTICATION

A process calling a remote procedure might need to identify itself as it is identified on the UNIX® system. The value of the credential's discriminant of an RPC call message is **AUTH_UNIX**. The bytes of the credential's opaque body encode the following structure:

```
struct auth_unix {
        unsigned        stamp;
        string          machinename;
        unsigned        uid;
        unsigned        gid;
        unsigned        gids;
};
```

The parameters in the structure are defined as follows:

stamp	Specifies the arbitrary ID generated by the caller's workstation.
machinename	Specifies the name of the caller's workstation. The name must not exceed 255 bytes in length.
uid	Specifies the caller's effective user ID.
gid	Specifies the caller's effective group ID.
gids	Specifies the counted array of group IDs that contain the caller as a member. A maximum of 10 groups is allowed.

TIPS

The verifier accompanying the credentials should be AUTH_NULL.

The value of the discriminant in the response verifier of the reply message from the server is either AUTH_NULL or AUTH_SHORT. If the value is AUTH_SHORT, the bytes of the response verifier's string encode an opaque structure. The new opaque structure can then be passed to the server in place of the original AUTH_UNIX credentials. The server maintains a cache that maps shorthand opaque structures (passed back by way of an AUTH_SHORT-style response verifier) to the original credentials of the caller. The caller saves network bandwidth and server CPU time when the shorthand credentials are used.

> **DO YOU KNOW**
>
> **The server can eliminate, or flush, the shorthand opaque structures at any time. If this happens, the RPC message will be rejected due to an AUTH_REJECTEDCRED authentication error. The original AUTH_UNIX credentials can be used when this happens.**

8.4.1. UNIX Authentication on the Client Side

When a caller creates a new RPC client handle, the authentication handle of the appropriate transport is set to the default by the **authnone_create** subroutine. The default for an RPC authentication handle is NULL. After creating the client handle, the client can select UNIX authentication with the **authunix_create** routine. This routine creates an authentication handle with operating system permissions and causes each remote procedure call associated with the handle to carry UNIX® credentials.

> **DO YOU KNOW**
>
> **Authentication information can be destroyed with the** auth_destroy **subroutine. Authentication information should be destroyed if one is attempting to conserve memory.**

8.4.2. UNIX Authentication on the Server Side

Dealing with authentication issues on the server side is more difficult than dealing with them on the client side. The caller's RPC package passes the service dispatch routine a request that has an arbitrary authentication style associated with it. The server must then determine which style of authentication the caller used and whether the style is supported by the RPC package.

If the authentication parameter type is not suitable for the calling service, the service dispatch routine calls the **svcerr_weakauth** routine to refuse the remote procedure call. It is *not* customary for the server to check the authentication parameters associated with procedure 0 (**NULLPROC**).

If the service does not have the requested protocol, the service dispatch returns a status for access denied. The **svcerr_systemerr** primitive is called to detect a system error that is not covered by a service protocol.

8.5. DATA ENCRYPTION STANDARD (DES) AUTHENTICATION

DES authentication offers more security features than UNIX® authentication. For DES authentication to work, the keyserv command must be running on both the server and client machines. The users at these workstations need public keys assigned in the public key database by the person administering the network. Additionally, each user's secret key must be decrypted using their keylogin command password.

DES authentication can handle the following UNIX problems:

- The naming scheme within UNIX authentication is UNIX system oriented.
- UNIX authentication lacks a verifier, thereby allowing falsification of credentials.

8.5.1. DES Authentication Naming Scheme

DES addresses the caller with a simple string of characters instead of an integer specific to a particular operating system. This string of characters is known as the caller's network name, or *net name*. The server is allowed to interpret the contents of the net name only to identify the caller. Therefore, net names should be unique for each caller in the network.

Each operating system is responsible for implementing DES authentication to generate unique net names for calling on remote servers. Because operating systems can already distinguish local users to their systems, extending this mechanism to the network is simple.

8.5.2. DES Authentication Verifiers

Unlike UNIX authentication, DES authentication has a verifier that permits the server to validate the client's credential and the client to validate the server's credential. The content of this verifier is primarily an encrypted time stamp. The time stamp is encrypted by the client and decrypted by the server. If the time stamp is close to real time, then the client encrypted it correctly. To encrypt the time stamp correctly, the client must have the conversation key of the RPC session. The client with the conversation key is the authentic client.

The *conversation key* is a DES key that the client generates and includes in its first remote procedure call to the server. The conversation key is encrypted using a public key scheme in the first transaction. The particular public key scheme used in DES authentication is the Diffie-Hellman system with 192-bit keys. For more information, see Diffie-Hellman Encryption.

For successful validation, both the client and the server need the same notion of current time. If network time synchronization cannot be guaranteed, the client can synchronize with the server before beginning the conversation, perhaps by consulting the Internet Time Server (TIME).

8.5.3. DES Authentication on the Server Side

The method for determining the validity of a client's time stamp depends on which transaction is under consideration.

For the first transaction, the server checks only that the time stamp has not expired. For subsequent transactions, the server verifies that the time stamp is greater than the previous time stamp from the same client, and that the time stamp has not expired. A time stamp has expired if the server's time is later than the sum of the client's time stamp plus the client's window. The sum of the time stamp plus the client's window can be thought of as the lifetime of the credential.

8.5.4. DES Authentication on the Client Side

In the first transaction to the server, the client sends an encrypted item, the *window verifier*, that must equal the client's window minus one, as an added check. Otherwise, the client could successfully send random data instead of the time stamp. Other values for the credential are rejected by the server. If the window verifier is accepted by the server, the server returns to the client a verifier equal to the encrypted time stamp, minus one second. If the client receives a different time stamp from the server, the client rejects it.

For subsequent transactions, the client's time stamp is valid if it is greater than the previous time stamp, and has not expired. A time stamp has expired if the server's time is later than the sum of the client's time stamp plus the client's window. The sum of the time stamp plus the client window can be thought of as the lifetime of the credential.

To use DES authentication, the programmer must set the client authentication handle using the **authdes_create** subroutine. This subroutine requires the network name of the owner of the server process, a lifetime for the credential, the address of the host with which to synchronize, and the address of a DES encryption key to use for encrypting time stamps and data.

8.5.5. Nicknames

The server's DES authentication subsystem returns a *nickname* to the client in the verifier response to the first transaction.

- The nickname is an unsigned integer.
- The nickname is likely to be an index into a table on the server that stores each client's net name, decrypted DES key, and window.
- The client can use the nickname in all subsequent transactions instead of passing its net name, encrypted DES key, and window each time.
- The nickname is not required, but it saves time.

8.6. CLOCK SYNCHRONIZATION

Although the client and server clocks are originally synchronized, they can lose this synchronization. When this happens, the client RPC subsystem normally receives the **RPC_AUTHERROR** error message and should resynchronize.

A client can receive the **RPC_AUTHERROR** message even when the clocks are synchronized. The message indicates that the server's nickname table has been flushed either because of the table's size limitations or a server crash. To receive new nicknames, all clients must resend their original credentials to the server.

DES Authentication Protocol

DES authentication has the following form of eXternal Data Representation (XDR) enumeration:

```
enum authdes_namekind {
        ADN_FULLNAME = 0,
        ADN_NICKNAME = 1
};
typedef opaque des_block[8];
const MAXNETNAMELEN = 255;
```

A credential is either a client's full network name or its nickname. For the first transaction with the server, the client must use its full name. For subsequent transactions, the client can use its nickname.

TIPS

DES authentication protocol includes a 64-bit block of encrypted DES data and specifies the maximum length of a network user's name.

The **authdes_cred union** provides a switch between the full-name and nickname forms, as follows:

```
union authdes_cred switch (authdes_namekind adc_namekind)
{
        case ADN_FULLNAME:
                authdes_fullname adc_fullname;
        case ADN_NICKNAME:
                unsigned int adc_nickname;
};
```

The full name contains the network name of the client, an encrypted conversation key, and the window. The window is actually a lifetime for the credential. The server can terminate a client's time stamp and not grant the request if the time indicated by the verifier time stamp plus the window has expired. In the first transaction, the server confirms that the window verifier is one second less than the window. To ensure that requests are granted only once, the server can require time stamps in subsequent requests to be greater than the client's previous time stamps.

The structure for a credential using the client's full network name follows:

```
struct authdes_fullname {
  string name<MAXNETNAMELEN>; /* name of client              */
 des_block key;            /*PK encrypted conversation key*/
   unsigned int window;  /* encrypted window              */
};
```

A time stamp encodes the time since midnight, January 1, 1970. The structure for the time stamp follows:

```
struct timestamp {
        unsigned int seconds;     /* seconds              */
        unsigned int useconds;  /* and microseconds   */
The client verifier has the following structure:
struct {
        adv_timestamp;             /* one DES block           */
        adc_fullname.window;       /* one half DES block      */
        adv_winverf;               /* one half DES block      */
}
```

The window verifier is only used in the first transaction. In conjunction with the **fullname** credential, these items are packed into the structure shown previously before being encrypted.

This structure is encrypted using CBC mode encryption with an input vector of 0. All other time stamp encryptions use ECB mode encryption. The client's verifier has the following structure:

```
struct authdes_verf_clnt {
    timestamp adv_timestamp;  /* encrypted timestamp      */
        unsigned int adv_winverf;     /* encrypted window
verifier */
};
```

The server returns the client's time stamp, minus one second, in an encrypted response verifier. This verifier also sends the client an unencrypted nickname to be used in future transactions. The verifier from the server has the following structure:

```
struct authdes_verf_svr {
   timestamp adv_timeverf;      /* encrypted verifier       */
  unsigned int adv_nickname;  /* new nickname for client */
};
```

8.7. DIFFIE-HELLMAN ENCRYPTION

Secure RPC authentication is based on the Diffie-Hellman *exponential key exchange* system. Each Secure RPC principal has a secret key and a public key, both of which are stored on the Secure RPC server. The public key is stored unencrypted; the secret key is stored encrypted with the principal's password. Both keys are typically hexadecimal numbers of several hundred digits.

The public key scheme used in DES authentication is Diffie-Hellman with 192-bit keys. The Diffie-Hellman encryption scheme includes two constants:

- BASE
- MODULUS.

Their values for these for the DES authentication protocol are:

```
const BASE = 3;
const MODULUS = "d4a0ba0250b6fd2ec6
        26e7efd637df76c716e22d0944b88b"; /* hex */
```

Two programmers, A and B, can send encrypted messages to each other in the following manner. First, programmers A and B independently generate secret keys at random, which can be represented as SK(A) and SK(B). Both programmers then publish their public keys PK(A) and PK(B) in a public directory. These public keys are computed from the secret keys as follows:

```
PK(A) = ( BASE ** SK(A) ) mod MODULUS
PK(B) = ( BASE ** SK(B) ) mod MODULUS
```

The ** (double asterisk) notation represents exponentiation. Programmers A and B can both arrive at the common key, represented here as CK(A, B), without revealing their secret keys.

Programmer A computes:

```
CK(A, B) = ( PK(B) ** SK(A)) mod MODULUS
```

while programmer B computes:

```
CK(A, B) = ( PK(A) ** SK(B)) mod MODULUS
```

These two can be shown to be equivalent:

```
(PK(B) ** SK(A)) mod MODULUS = (PK(A) ** SK(B)) mod MODULUS
```

If the mod MODULUS parameter is omitted, modulo arithmetic can simplify things as follows:

```
PK(B) ** SK(A) = PK(A) ** SK(B)
```

Then, if the result of the previous computation on B replaces PK(B) and the previous computation of A replaces PK(A), the equation is:

```
((BASE ** SK(B)) ** SK(A) = (BASE ** SK(A)) ** SK(B)
```

This equation can be simplified as follows:

```
BASE ** (SK(A) * SK(B)) = BASE ** (SK(A) * SK(B))
```

This produces a common key CK(A, B). This common key is not used directly to encrypt the time stamps used in the protocol. Instead, it is used to encrypt a conversation key that is then used to encrypt the time stamps. In this way, the common key is used as little as possible to prevent it from being broken.

Breaking the conversation key usually has less serious consequences because conversations are relatively shortlived.

The conversation key is encrypted using 56-bit DES keys, while the common key is 192 bits. To reduce the number of bits, 56 bits are selected from the common key as follows. The middle eight bytes are selected from the common key and parity is added to the lower order bit of each byte, producing a 56-bit key with eight bits of parity

8.8. LIMITATIONS OF SECURE RPC

➢ **Every network client must be individually modified for use with Secure RPC.** Although Secure RPC is a transparent modification to Sun's underlying RPC system, the current design of Sun's RPC library requires an application program to specify individually which authentication system (AUTH_NONE, AUTH_UNIX, AUTH_DES, or AUTH_KERB) the program wants to use. For this reason, every client that uses a network service must be individually modified to use AUTH_DES authentication.

  Although the modifications required are trivial, a better approach would be to allow the user to specify the authentication service requested in an environment variable, or on some other per-user or per-site, rather than per-program, basis.

- **There is a performance penalty.** Secure RPC penalizes every RPC transaction that uses it, because the RPC authenticator must be decrypted using DES to verify each transmission. Fortunately, the performance penalty is small: On a Sun-4, only 1.5 milliseconds are required for the decryption. In comparison, the time to complete an average NFS transaction is about 20 milliseconds, making the performance penalty about eight percent.
- **Secure RPC does not provide for data integrity or confidentiality**. Secure RPC authenticates the user, but it does not protect the data that is transmitted with either encryption or digital signatures. It is the responsibility of programs using Secure RPC to encrypt using a suitable key and algorithm.
- **It may be possible to break the public key.** Any piece of information encrypted with the Diffie-Hellman public key encryption system used in Secure RPC can be decrypted if an attacker can calculate the discrete logarithm of the public key. Secure RPC makes the public key and the encrypted secret key available to RPC client computers on the network. Thus, keys that are secure today may be broken tomorrow.
- **It may be possible to break the secret key.** The Secure RPC secret key is encrypted with a 56-bit DES key and is made publicly available on the network server. As computers become faster, the possibility of a brute force attack against the user's encrypted secret key may become a reality.

Network Information Service (NIS) is a distributed database system that lets many computers share password files, group files, host tables, and other files over the network. Although the files appear to be available on every computer, they are actually stored on only a single computer, called the NIS *master server* (and possibly replicated on a backup, or *slave* server). The other computers on the network, *NIS clients*, can use the databases stored on the master server (like */etc/passwd)* as if they were stored locally. These databases are called NIS *maps*.NIS+ is advanced version of NIS.

> **DO YOU KNOW**
>
> **Secure RPC is clearly better than plain RPC. Unfortunately, because Secure RPC requires the use of either NIS or NIS+, some multi-vendor sites have chosen not to use it**

8.9. KERBEROS AUTHENTICATION

Kerberos is a computer network authentication protocol, which allows nodes communicating over a non-secure network to prove their identity to one another in a secure manner. Its designers aimed primarily at a client–server model, and it provides mutual authentication — both the user and the server verify each other's identity.

Kerberos builds on symmetric key cryptography and requires a trusted third party, and optionally may use public-key cryptography by utilizing asymmetric key cryptography during certain phases of authentication.

Kerberos authentication is based entirely on the knowledge of passwords that are stored on the Kerberos Server. Unlike UNIX passwords, which are encrypted with a one-way algorithm that cannot be reversed, Kerberos passwords are stored on the server encrypted with a conventional encryption algorithm - in this case, DES - so that they can be decrypted by the server when needed. A user proves her identity to the Kerberos Server by demonstrating knowledge of her key.

8.9.1. Kerberos vs. Secure RPC

Kerberos is an authentication system, not an RPC system. Kerberos can be used with a variety of RPC schemes: versions of Kerberos are available for Sun RPC and for the X Window System. But Kerberos can also be used simply for exchanging keys. For example, there is a version of the telnet command that uses Kerberos to exchange a cryptographic key. MIT also modified the NFS protocol to work with Kerberos; the modification was a simple kernel patch to the NFS server that maintained a mapping between authenticated users on discrete hosts and UIDS on the NFS server.

There are other important differences between Kerberos and Secure RPC:

➢ Secure RPC passwords are based on public key cryptography, not secret key cryptography. When the user wishes to prove her identity to the NIS server, she encrypts an authenticator and sends it to the Secure RPC server, which then decrypts her authenticator using the user's widely available public key. With Kerberos, identity is provided by knowing the secret key.

➢ Secure RPC stores both the user's secret key and public key on the NIS server. The secret key is encrypted with the user's password and made available to the network, but the network does not have the ability to decrypt it. Thus, with Secure RPC, there is no need for a specially secured "authentication server" to establish the identity of users on the network.

➢ Secure RPC is built into Sun's RPC system. While Kerberos requires that each application be specifically tailored. Secure RPC is a transparent modification to Sun's low-level RPC that works with any RPC-based service. Any application can use it simply by requesting AUTH_DES authentication.

8.9.2. Kerberos Limitations

Although Kerberos is an excellent solution to a difficult problem, it has several shortcomings:

➢ **Every network service must be individually modified for use with Kerberos.** Because of the Kerberos design, every program that uses Kerberos must be modified. The process of performing these modifications is often called "Kerberizing" the application. The amount of work that this entails depends entirely on the application program. Of course, to Kerberize an application, you must have the application's source code.

➢ **Kerberos doesn't work well in a time-sharing environment.** Kerberos is designed for an environment in which there is one user per workstation. Because of the difficulty of sharing data between different processes running on the same UNIX computer, Kerberos keeps tickets in the /tmp directory. If a user is sharing the computer with several other people, it is possible that the user's tickets can be stolen, that is, copied by an attacker. Stolen tickets can then be used to obtain fraudulent service.

➢ **Kerberos requires a secure Kerberos Server.** By design, Kerberos requires that there be a secure central server that maintains the master password database. To ensure security, a site should use the Kerberos Server for absolutely nothing beyond running the Kerberos Server program. The Kerberos Server must be kept under lock and key, in a physically secure area. In some environments, maintaining such a server is an administrative and/or financial burden.

- **Kerberos requires a continuously available Kerberos Server.** If the Kerberos Server goes down, the Kerberos network is unusable.
- **Kerberos stores all passwords encrypted with a single key**. Adding to the difficulty of running a secure server is the fact that the Kerberos Server stores all passwords encrypted with the server's master key, which happens to be located on the same hard disk as the encrypted passwords. This means that, in the event the Kerberos Server is compromised, all user passwords must be changed.
- **Kerberos does not protect against modifications to system software (Trojan horses)**. Kerberos does not have the computer authenticate itself to the user - that is, there is no way for a user sitting at a computer to determine whether the computer has been compromised. This failing is easily exploited by a knowledgeable attacker.
- An intruder, for example, can modify the workstation's system software so every username/password combination typed is recorded automatically or sent electronically to another machine controlled by the attacker. Alternatively, a malicious attacker can simply modify the workstation's software to spuriously delete the user's files after the user has logged in and authenticated himself to the File Server Service. Both of these problems are consequences of the fact that, even in a networked environment, many workstations (including Project Athena's) contain local copies of the programs that they run.
- **Kerberos may result in a cascading loss of trust.** Another problem with Kerberos is that if a server password or a user password is broken or otherwise disclosed, it is possible for an eavesdropper to use that password to decrypt other tickets and use this information to spoof servers and users.

Kerberos is a workable system for network security, and it is still widely used. But more importantly, the principles behind Kerberos are increasingly available in network security systems that are available directly from vendors.

8.10. RPC PORT MAPPER PROGRAM

Client programs must find the port numbers of the server programs that they intend to use. Network transports do not provide such a service; they merely provide process-to-process message transfer across a network. A message typically contains a transport address consisting of a network number, a host number, and a port number.

A *port* is a logical communications channel in a host. A server process receives messages from the network by waiting on a port. How a process waits on a port varies from one operating system to another, but all systems provide mechanisms that suspend processes until a message arrives at a port. Therefore, messages are sent to the ports at which receiving processes wait for messages.

Ports allow message receivers to be specified in a way that is independent of the conventions of the receiving operating system. The port mapper protocol defines a network service that permits clients to look up the port number of any remote program supported by the server. Because the port mapper program can be implemented on any transport that provides the equivalent of ports, it works for all clients, all servers, and all networks.

The port mapper program maps Remote Procedure Call (RPC) program and version numbers to transport-specific port numbers. The port mapper program makes dynamic binding of remote programs possible. This is desirable because the range of reserved port numbers is small and the number of potential remote programs large. When running only the port mapper on a reserved port, the port numbers of other remote programs can be determined by querying the port mapper.

The port mapper also aids in broadcast RPC. A given RPC program usually has different port number bindings on different machines, so there is no way to directly broadcast to all of these programs. The port mapper, however, has a fixed port number. To broadcast to a given program, the client sends its message to the port mapper located at the broadcast address. Each port mapper that picks up the broadcast then calls the local service specified by the client. When the port mapper receives a reply from the local service, it sends the reply back to the client.

8.10.1. Registering Ports

Every port mapper on every host is associated with port number 111. The port mapper is the only network service that must have a dedicated port. Other network services can be assigned port numbers either statically or dynamically, as long as the services register their ports with their host's port mapper. Typically, a server program based on an RPC library gets a port number at run time by calling an RPC library procedure.

DO YOU KNOW

Service on a host can be associated with a different port every time its server program is started. For example, a given network service can be associated with port number 256 on one server and port number 885 on another.

The delegation of port-to-remote program mapping to a port mapper also automates port number administration. Statically mapping ports and remote programs in a file duplicated on each client requires updating all mapping files whenever a new remote program is introduced to a network. The alternative solution, placing the port-to-program mappings in a shared Network File System (NFS) file, would be too centralized. If the file server were to go down in this case, the entire network would also.

The port-to-program mappings, which are maintained by the port mapper server, are called a *portmap*. The port mapper is started automatically whenever a machine is booted. Both the server programs and the client programs call port mapper procedures. As part of its initialization, a server program calls its host's port mapper to create a portmap entry. Whereas server programs call port mapper programs to update portmap entries, clients call port mapper programs to query portmap entries. To find a remote program's port, a client sends an RPC call message to a server's port mapper. If the remote program is supported on the server, the port mapper returns the relevant port number in an RPC reply message. The client program can then send RPC call messages to the remote program's port. A client program can minimize port mapper calls by caching the port numbers of recently called remote programs.

TIPS

The port mapper provides an inherently useful service because a portmap is a set of associations between registrants and ports.

Port Mapper Procedures

The port mapper program currently supports two protocols: User Datagram Protocol (UDP) and Transmission Control Protocol/Internet Protocol (TCP/IP). The port mapper is contacted by port number 111 on both protocols.

A description of the port mapper procedures follows.

PMAPPROC_NULL

This procedure does no work. By convention, procedure 0 of any protocol takes no parameters and returns no results.

PMAPPROC_SET

When a program first becomes available on a machine, it registers itself with the port mapper program on that machine. The program passes its program number (*prog*), version number (*vers*), transport protocol number (*prot*), and the port (*port*) on which it awaits service request. The procedure returns a Boolean response whose value is either True if the procedure successfully established the mapping, or False if otherwise. The procedure does not establish a mapping if the values for the *prog*, *vers*, and *prot* parameters indicate a mapping already exists.

PMAPPROC_UNSET

When a program becomes unavailable, it should unregister itself with the port mapper program on the same machine. The parameters and results have meanings identical to those of the PMAPPROC_SET procedure. The protocol and port number fields of the argument are ignored.

PMAPPROC_GETPORT

Given a program number (*prog*), version number (*vers*), and transport protocol number (*prot*), this procedure returns the port number on which the program is awaiting call requests. A *port* value of zero means the program has not been registered. The *port* parameter of the argument is then ignored.

PMAPPROC_DUMP

This procedure enumerates all entries in the port mapper database. The procedure takes no parameters and returns a list of *prog*, *vers*, *prot*, and *port* values.

PMAPPROC_CALLIT

This procedure allows a caller to call another remote procedure on the same machine without knowing the remote procedure's port number. It supports broadcasts to arbitrary remote programs through the well-known port mapper port. The *prog*, *vers*, and *proc* parameters, and the bytes of the *args* parameter of an RPC call represent the program number, version number, procedure number, and arguments, respectively. The PMAPPROC_CALLIT procedure sends a response only if the procedure is successfully run. The port mapper communicates with the remote program using UDP only. The procedure returns the remote program's port number, and the bytes of results are the results of the remote procedure.

8.11. RPC LANGUAGE

The Remote Procedure Call Language (RPCL) is identical to the eXternal Data Representation (XDR) language, except for the added program definition.

8.11.1. RPC Language Descriptions

Because XDR data types are described in a formal language, procedures that operate on these data types must be described in a formal language. The RPCL, an extension to the XDR language, is used for this purpose.

RPC uses RPCL as the input language for its protocol and routines. RPCL specifies data types used by RPC and generates XDR routines that standardize representation of the types. To implement service protocols and routines, RPCL uses the **rpcgen** command to compile input in corresponding C language code.

RPC language descriptions include:

- ➢ Definitions
- ➢ Structures
- ➢ Unions
- ➢ Enumerations
- ➢ Type Definitions
- ➢ Constants
- ➢ Programs
- ➢ Declarations

An RPCL file consists of a series of definitions in the following format:

```
definition-list:
      definition ";"
      definition ";" definition-list
```

RPCL recognizes the following six types of definitions:

```
definition:
      enum-definition
      struct-definition
      union-definition
      typedef-definition
      const-definition
      program-definition
```

8.11.2. Structures

The C language structures are usually located in header files in either the **/usr/include** or **/usr/include/sys** directory, but they can be located in any directory in the file system. An XDR structure is declared almost exactly like its C language counterpart; for example:

```
struct-definition:
        "struct" struct-ident "{"
        declaration-list
        "}"
declaration-list:
        declaration ";"
        declaration ";" declaration-list
```

Compare the following XDR structure to a two-dimensional coordinate with the C structure that it is compiled into in the output header file.

```
struct coord {                  struct coord {
        int x;          -->             int x;
        int y;                          int y;
};                              };
                                typedef struct coord coord;
```

Here, the output is identical to the input, except for the added typedef at the end of the output. As a result, the programmer can use coord instead of struct coord when declaring items.

8.11.3. Unions

XDR unions are discriminated unions and look different from C unions. XDR unions are more analogous to Pascal variant records than to C unions. Following is an XDR union definition:

```
union-definition:
        "union" union-ident "switch" "(" declaration ")" "{"
                case-list
        "}"
case-list:
        "case" value ":" declaration ";"
        "default" ":" declaration ";"
        "case" value ":" declaration ";" case-list
```

Following is an example of a type that might be returned as the result of a read data operation. If there is no error, the type returns a block of data; otherwise, it returns nothing.

```
union read_result switch (int errno) {
case 0
        opaque data[1024];
default:
```

```
        void;
    };
```

The type is compiled into the following structure:

```
struct read_result {
        int errno;
        union {
                char data[1024];
        }read_result_u;
};

typedef struct read_result read_result;
```

Note : The union component of this output structure is identical to the type, except for the trailing _u.

8.11.4. Enumerations

XDR enumerations have the same syntax as C enumerations.

```
enum-definition:
        "enum" enum-ident "{"
        enum-value-list
        "}"
enum-value-list:
        enum-value
        enum-value "," enum-value-list
enum-value:
        enum-value-ident
        enum-value-ident "=" value
```

Compare the following example of an XDR enumeration with the C enumeration it is compiled into.

```
enum colortype {              enum colortype {
    RED = 0,                          RED = 0,
    GREEN = 1,        -->             GREEN = 1,
    BLUE = 2                          BLUE = 2,
};                            };
                              typedef enum colortype colortype;
```

8.11.5. Type Definitions

XDR type definitions (typedefs) have the same syntax as C typedefs.

```
typedef-definition:
        "typedef" declaration
```

The following example defines an fname_type used for declaring file-name strings with a maximum length of 255 characters.

```
typedef string fname_type<255>; -> typedef char *fname_type;
```

8.11.6. Constants

XDR constants can be used wherever an integer constant is required. The definition for a constant is:

```
const-definition:
        "const" const-ident "=" integer
```

For example, the following defines a constant DOZEN equal to 12.

```
const DOZEN = 12; -> #define DOZEN 12
```

8.11.7. Programs

RPC programs are declared using the following syntax:

```
program-definition:
        "program" program-ident "{"
               version-list
        "}" "=" value
version-list:
        version ";"
        version ";" version-list
version:
        "version" version-ident "{"
               procedure-list
        "}" "=" value
procedure-list:
        procedure ";"
        procedure ";" procedure-list
procedure:
        type-ident procedure-ident "(" type-ident ")" "=" value
```

The time protocol is defined as follows:

```
/*
 * time.x: Get or set the time. Time is represented as number
```

```
  * of seconds since 0:00, January 1, 1970.
  */
program TIMEPROG {
     version TIMEVERS {
          unsigned int TIMEGET (void) = 1;
          void TIMESET (unsigned) = 2;
     } = 1;
} = 44;
```

This file compiles into the following #define statements in the output header file:

```
#define TIMEPROG 44
#define TIMEVERS 1
#define TIMEGET 1
#define TIMESET 2
```

8.11.8. Declarations

XDR includes four types of declarations: simple declarations, fixed-length array declarations, variable-length array declarations, and pointer declarations. These declarations have the following forms:

```
declaration:
      simple-declaration
      fixed-array-declaration
      variable-array-declaration
      pointer-declaration
```

8.11.8.1. Simple Declarations

Simple XDR declarations are like simple C declarations, as follows:

simple-declaration:

```
      type-ident variable-ident
```

An example of a simple declaration is:

```
colortype color;   ->  colortype color;
```

8.11.8.2. Fixed-length Array Declarations

Fixed-length array declarations are like C array declarations, as follows:

```
fixed-array-declaration:

      type-ident variable-ident "[" value "]"
```

An example of a fixed-length array declaration is:

```
colortype palette[8];   ->  colortype palette[8]
```

8.11.8.3. Variable-length Array Declarations

Variable-length array declarations have no explicit syntax in C, so XDR invents its own syntax using angle brackets. The maximum size is specified between the angle brackets. A specific size can be omitted to indicate that the array may be of any size.

```
variable-array-declaration:
      type-ident variable-ident "<" value ">"
      type-ident variable-ident "<" ">"
```

An example of a set of variable-length array declarations is:

```
int heights<12>;             /* at most 12 items     */
int widths<>;                /* any number of items  */
```

TIPS

The maximum size is specified between the angle brackets. The number, but not the angle brackets, may be omitted to indicate that the array can be of any size.

Because variable-length arrays have no explicit syntax in C, these declarations are actually compiled into structure definitions, signified by struct. For example, the heights declaration is compiled into the following structure:

```
struct {
     u_int heights_len;    /* # of items in array */
     int *heights_val;     /* # pointer to array */
} heights;
```

8.11.9. Pointer Declarations

Pointer declarations are made in XDR exactly as they are in C. The programmer cannot send pointers over a network, but can use XDR pointers for sending recursive data types such as lists and trees. In XDR language, the type is called optional-data, instead of pointer. Pointer declarations have the following form in XDR language:

```
pointer-declaration:
      type-ident "*" variable-ident
```

An example of a pointer declaration is:

```
listitem *next;  ->  listitem *next;
```

8.12. RPCL SYNTAX REQUIREMENTS FOR PROGRAM DEFINITION

The RPCL has the following syntax requirements:

- The **program** and **version** keywords are added and cannot be used as identifiers.
- A version name cannot occur more than once within the scope of a program definition. Nor can a version number occur more than once within the scope of a program definition.
- A procedure name cannot occur more than once within the scope of a version definition. Nor can a procedure number occur more than once within the scope of a version definition.
- Program identifiers are in the same name space as the **constant** and **type** identifiers.
- Only unsigned constants can be assigned to **program**, **version**, and **procedure** definitions.

8.13. EXCEPTIONS TO THE RPCL RULES

Exceptions to the RPC language rules include Booleans, strings, opaque data, and voids.

8.13.1. Booleans

The C language has no built-in Boolean type. However, the RPC library uses a Boolean type called bool_t, which is either True or False. Objects that are declared as type bool in XDR language are compiled into bool_t in the output header file; for example:

```
bool married;    ->   bool_t married;
```

8.13.2. Strings

The C language has no built-in string type. Instead, it uses the null-terminated char * convention. In XDR language, strings are declared using the **string** keyword, and then compiled into char * in the output header file. The maximum size contained in the angle brackets specifies the maximum number of characters allowed in the strings (not counting the null character). The maximum size may be left off, indicating a string of arbitrary length.

Compare the following examples:

```
string name<32>;    ->   char *name;
string longname<>;    ->   char *longname;
```

8.13.3. Opaque Data

Opaque data is used in RPC and XDR to describe untyped data, which consists of sequences of arbitrary bytes. Opaque data may be declared either as a fixed-length or variable-length array, as in the following examples:

```
opaque diskblock[512];    ->   char diskblock[512];
opaque filedata<1024>;    ->   struct {
                                      u_int filedata_len;
                                      char *filedata_val;
                                 } filedata
```

8.13.4. Voids

In a void declaration, the variable is not named. The declaration is void. Void declarations can occur as the argument or result of a remote procedure in only two places: union definitions and program definitions.

8.14. RPCGEN PROTOCOL COMPILER

The **rpcgen** protocol compiler accepts a remote program interface definition written in the Remote Procedure Call language (RPCL), which is similar to the C language. The **rpcgen** compiler helps programmers write RPC applications in a simple and direct manner. The **rpcgen** compiler debugs the network interface code, thereby allowing programmers to spend their time debugging the main features of their applications.

The **rpcgen** compiler produces a C language output that includes the following:

- Stub versions of the client and server routines
- Server skeleton
- eXternal Data Representation (XDR) filter routines for parameters and results
- A header file that contains common definitions of constants and macros

Client stubs interface with the RPC library to effectively hide the network from its callers. Server stubs similarly hide the network from server procedures invoked by remote clients. The **rpcgen** output files can be compiled and linked in the usual way. Using any language, programmers write server procedures and link them with the server skeleton to get an executable server program.

When application programs use the **rpcgen** compiler, there are many details to consider. Of particular importance is the writing of XDR routines needed to convert procedure arguments and results into the network format, and vice versa.

HIGHLIGHTS

1. Secure RPC is based on :
   - Public key cryptography
   - Secret key cryptography
2. NULL authentication is used when RPC caller does known its own identity or the server does not need to know the caller's identity.
3. DES authentication provides better security than UNIX.
4. Secure RPC authentication is based on the Diffie-Hellman exponential key exchange system.
5. Limitations of secure RPC :
   - There is performance penalty
   - Secure RPC does not provide for data integrity
   - It may be possible to break the public key.
   - It may be possible to break the secret key.

6. Authentication is of following types :
   - Null authentication.
   - UNIX authentication.
   - DES authentication
   - Kerberos
   - Diffie Hellman authentication
7. Kerberos authentication is based on knowledge of password.
8. Kerberos limitations
   - Not work in time sharing environment.
   - Require secure kerboros server.
9. RPC is based on language called RPCL
10. RPCL supports almost all the concept of RPC.

REVIEW QUESTIONS

1. What is RPC authentication.
2. What is secure RPC ?
3. What is NULL authentication.
4. What is Unix authentication.
5. Explain DES authentication.
6. Explain Diffie-Hellman encryption.
7. What are the limitations of secure RPC ?
8. What is kerberos authentication.
9. What are the limitations of kerberos.
10. Explain features of RPC language.

❑❑❑

UNIT-V

9. NFS (NETWORK FILE SYSTEM)

NFS (Network File Systems)

Chapter 9

INSIDE THIS CHAPTER

9.1. INTRODUCTION [MDU May 2008, Dec. 2007; May 2006, Dec. 2006]

Using NFS, clients can mount partitions of a server as if they were physically connected to the client. In addition to simply allowing remote access to files over the network, NFS allows many (relatively) low-cost computer systems to share the same high-capacity disk drive at the same time. NFS server programs have been written for many different operating systems, which let users on UNIX workstations have remote access to files stored on a variety of different platforms

NFS is a technology for sharing resources between devices on a local area network (LAN). NFS allows data to be stored on central servers and easily accessed from client devices in a client/server network configuration via a process called *mounting*.

OR

NFS allows a system to share directories and files with others over a network. By using NFS, users and programs can access files on remote systems almost as if they were local file

OR

The NFS environment is a service that enables computers of different architectures running different operating systems to share file systems across a network

> **DO YOU KNOW**
>
> **NFS was developed by sun microsystems in 1984.**

9.1.1. Advantages of NFS

(i) Data accessed by all users can be kept on a central host, with clients mounting this directory at boot time. For example, you can keep all user accounts on one host, and have all hosts on your network mount /home from that host. If installed alongside with NIS, users can then log into any system, and still work on one set of files.

(ii) Administrative data may be kept on a single host. No need to use rcp anymore to install the same stupid file on 20 different machines.

(iii) Local workstations use less disk space because commonly used data can be stored on a single machine and still remain accessible to others over the network.

(iv) There is no need for users to have separate home directories on every network machine. Home directories could be set up on the NFS server and made available throughout the network.

(v) Storage devices such as floppy disks, CDROM drives, and Zip® drives can be used by other machines on the network. This may reduce the number of removable media drives throughout the network.

9.1.2. Disadvantages of NFS

(i) It gives the right for unwanted people to access your hard drive over the network and thereby possibly read your email and delete all your files as well as break into your system)if you set it up incorrectly.

(ii) NFS is built on top of Sun's RPC (Remote Procedure Call), and in most cases uses RPC for user authentication. Unless a secure form of RPC is used, NFS can be easily spoofed.

(iii) Even when Secure RPC is used, information sent by NFS over the network is not encrypted, and is thus subject to monitoring and eavesdropping. As we mention elsewhere, the data can be intercepted and replaced (thereby corrupting or Trojaning files being imported via NFS).

(iv) NFS uses the standard UNIX filesystem for access control, opening the networked filesystem to many of the same problems as a local filesystem

DO YOU KNOW

Security is biggest challenge to NFS.

9.2. VARIOUS VERSION OF NFS

1. NFSv1-Sun used version 1 only for in-house experimental purposes. When the development team added substantial changes to NFS version 1 and released it outside of Sun, they decided to release the new version as v2

2. NFSv2- includes performance improvements, mandates strong security, and introduces a stateful protocol.

During its 10-year life, many subtle, undocumented changes were made to the NFS version 2 specification. Some vendors allowed NFS version 2 to read or write more than 4K bytes at a time; others increased the number of groups provided as part of the RPC authentication from 8 to 16. Although these minor changes created occasional incompatibilities between different NFS implementations, NFS version 2 provided a remarkable degree of compatibility between systems made by different vendors.

3. NFSv3 .

- Support for 64-bit file sizes and offsets, to handle files larger than 2 gigabytes (GB);
- Support for asynchronous writes on the server, to improve write performance;
- Additional file attributes in many replies, to avoid the need to re-fetch them;
- A READDIRPLUS operation, to get file handles and attributes along with file names when scanning a directory;
- Assorted other improvements

4. NFSv4 : Includes performance improvements, mandates strong security, and introduces a stateful protocol. Version 4 became the first version developed with the Internet Engineering Task Force (IETF) after Sun Microsystems handed over the development of the NFS protocols.

TIPS

NFS Version 4 is the latest protocol specification for NFS and is defined in RFC 3530.

9.2.1. Difference between NFSv2 and NFSv3

- Version 2 of the NFS protocol limited file offsets to a 32-bit quantity, which limited the size of files accessible by clients to 4.2 GB. For users who regularly gain access to larger files, this was a severe limitation. NFS version 3 extended the file offsets and a number of other fields to 64-bits.
- NFS Version 2 limited the data transfer size to 8 KB. No single read or write request could exceed 8 KB. This limits performance on high-bandwidth networks because it artificially increases the number of NFS requests to transfer a given amount of data. NFS version 3 removed that limitation and allows the client and server to negotiate a maximum transfer size.
- Version 2 NFS servers must commit data written by a client to stable storage before responding affirmatively to the client.
- NFS version 3 provides a new COMMIT operation that allows a client to perform unstable writes to a server followed by a COMMIT request. The server is required to verify that client data is on stable storage only when it receives the COMMIT operation. A mechanism is provided that allows the client to detect server loss of uncommitted data and recover.

DO YOU KNOW

NFS Version 2 is documented in RFC 1094 and was published in March 1989. NFS Version 3 is documented in RFC 1813 and was published in June 1995.

9.2.2. Difference between NFSv3 and NFSv4

Currently, there are three versions of NFS. NFS version 2 (NFSv2) is older and is widely supported. NFS version 3 (NFSv3) has more features, including 64bit file handles, Safe Async writes and more robust error handling. NFS version 4 (NFSv4) works through firewalls and on the Internet, no longer requires portmapper, supports ACLs, and utilizes stateful operations. Red Hat Enterprise Linux supports NFSv2, NFSv3, and NFSv4 clients, and when mounting a file system via NFS, Red Hat Enterprise Linux uses NFSv3 by default, if the server supports it.

All versions of NFS can use *Transmission Control Protocol* (*TCP*) running over an IP network, with NFSv4 requiring it. NFSv2 and NFSv3 can use the *User Datagram Protocol* (*UDP*) running over an IP network to provide a stateless network connection between the client and server.

When using NFSv2 or NFSv3 with UDP, the stateless UDP connection under normal conditions has less Protocol overhead than TCP which can translate into better performance on very clean, non-congested networks. The NFS server sends the client a file handle after the client is authorized to access the shared volume. This file handle is an opaque object stored on the server's side and is passed along with RPC requests from the client. The NFS server can be restarted without affecting the clients and the cookie remains intact. However, because UDP is stateless, if the server goes down unexpectedly, UDP clients continue to saturate the network with requests for the server. For this reason, TCP is the preferred protocol when connecting to an NFS server.

NFSv4 has no interaction with portmapper, rpc.mountd, rpc.lockd, and rpc.statd, since protocol support has been incorporated into the v4 protocol. NFSv4 listens on the well known TCP port (2049) which eliminates the need for the portmapper interaction. The mounting and locking protocols have been incorpated into the V4 protocol which eliminates the need for interaction with rpc.mountd and rpc.lockd.

DO YOU KNOW

TCP is the default transport protocol for NFS under Red Hat Enterprise Linux. UDP can be used for compatibility purposes as needed, but is not recommended for wide us-

NFS is based on two similar but distinct protocols:

- MOUNT and
- NFS.

Both make use of a data object known as a *file handle*. There is also a distributed protocol for file locking, which is not technically part of NFS, and which does not have any obvious security ramifications.

9.3. FILE HANDLES

Each object on the NFS-mounted filesystem is referenced by a unique object called a *file handle*. A file handle is viewed by the client as being *opaque* - the client cannot interpret the contents. However, to the server, the contents have considerable meaning. The file handles uniquely identify every file and directory on the server computer.

Under UNIX, a file handle consists of at least three important elements: the

- filesystem identifier,
- file identifier
- *generation count*.

The file identifier can be something as simple as an inode number to refer to a particular item on a partition. The filesystem identifier refers to the partition containing the file (inode numbers are unique per partition, but not per system). The file handle doesn't include a pathname; a pathname is not necessary and is in fact subject to change while a file is being accessed.

The generation count is a number that is incremented each time a file is unlinked and recreated. The generation count ensures that when a client references a file on the server, that file is in fact the same file that the server thinks it is. Without a generation count, two clients accessing the same file on the server could produce erroneous results if one client deleted the file and created a new file with the same inode number. The generation count prevents such situations from occurring: when the file is recreated, the generation number is incremented, and the second client gets an error message when it attempts to access the older, now nonexistent, file.

> **DO YOU KNOW**
>
> **Some NFS servers ignore the generation count in the file handle. These versions of NFS are considerably less secure, as they enable an attacker to easily create valid file handles for directories on the server.**

9.4. MOUNT PROTOCOL

The MOUNT protocol is used for the initial negotiation between the NFS client and the NFS server. Using MOUNT, a client can determine which filesystems are available for mounting and can obtain a token (the file handle) which is used to access the root directory of a particular filesystem. After that file handle is returned, it can thereafter be used to retrieve file handles for other directories and files on the server.

> **DO YOU KNOW**
>
> **The MOUNT protocol is based on Sun Micro-system's Remote Procedure Call (RPC) and External Data Representation (XDR) protocols.**

Another benefit of the MOUNT protocol is that you can export only a portion of a local partition to a remote client. By specifying that the root is a directory on the partition, the MOUNT service will return its file handle to the client. To the client, this file handle behaves exactly as one for the root of a partition: reads, writes, and directory lookups all behave the same way.

MOUNT is an RPC service. The service is provided by the mountd or rpc.mountd command, which is started automatically at boot.

The standard mountd can respond to seven different requests:

NULL Does nothing.

MNT Returns a file handle for a filesystem. Advises the mount daemon that a client has mounted the filesystem.

DUMP Returns the list of mounted filesystems.

UMNT Removes the mount entry for this client for a particular filesystem.

UMNTALL Removes all mount entries for this client.

EXPORT Returns the server's export list to the client.

> **TIPS**
>
> **Although the MOUNT protocol provides useful information within an organization, the information that it provides could be used by those outside an organization to launch an attack. For this reason, you should prevent people outside your organization from accessing your computer's mount daemon.**

Two ways of providing this protection are via the

- ➢ portmapper wrapper
- ➢ organizational firewall.

9.5. NFS PROTOCOL

The NFS protocol takes over where the MOUNT protocol leaves off. With the NFS protocol, a client can list the contents of an exported filesystem's directories; obtain file handles for other directories and files; and even create, read, or modify files.

Following is a list of the RPC functions that perform operations on directories:

CREATE	Creates (or truncates) a file in the directory
LINK	Creates a hard link
LOOKUP	Looks up a file in the directory
MKDIR	Makes a directory
READADDR	Reads the contents of a directory
REMOVE	Removes a file in the directory
RENAME	Renames a file in the directory
RMDIR	Removes a directory
SYMLINK	Creates a symbolic link

These RPC functions can be used with files:

GETATTR	Gets a file's attributes (owner, length, etc.)
SETATTR	Sets some of a file's attributes
READLINK	Reads a symbolic link's path
READ	Reads from a file.
WRITE	Writes to a file.

NFS version 3 adds a number of additional RPC functions. With the exception of MKNOD3, these new functions simply allow improved performance:

ACCESS	Determines if a user has the permission to access a particular file or directory.
FSINFO	Returns static information about a filesystem.
FSSTAT	Returns dynamic information about a filesystem.
MKNOD	Creates a device or special file on the remote filesystem.
READDIRPLUS	Reads a directory and returns the file attributes for each entry in the directory.
PATHCONF	Returns the attributes of a file specified by pathname.
COMMIT	Commits the NFS write cache to disk.

All communication between the NFS client and the NFS server is based upon Sun's RPC system, which lets programs running on one computer call subroutines that are executed on another.

RPC uses Sun's XDR system to allow the exchange of information between different kinds of computers.

DO YOU KNOW

For speed and simplicity, Sun built NFS upon the Internet User Datagram Protocol (UDP); however, NFS version 3 allows the use of TCP, which actually improves performance over low-bandwidth, high-latency links such as modem-based PPP connections.

How NFS Creates a Reliable File System from a Best-Effort Protocol

UDP is fast but only best-effort: "Best effort" means that the protocol does not guarantee that UDP packets transmitted will ever be delivered, or that they will be delivered in order. NFS works around this problem by requiring the NFS server to acknowledge every RPC command with a result code that indicates whether the command was successfully completed or not. If the NFS client does not get an acknowledgment within a certain amount of time, it retransmits the original command.

If the NFS client does not receive an acknowledgment, then UDP lost either the original RPC command or the RPC acknowledgment. If the original RPC command was lost, there is no problem - the server sees it for the first time when it is retransmitted. But if the acknowledgment was lost, the server will actually get the same NFS command twice.

For most NFS commands, this duplication of requests presents no problem. With READ, for example, the same block of data can be read once or a dozen times, without consequence. Even with the WRITE command, the same block of data can be written twice to the same point in the file, without consequence.

This is precisely the reason that NFS does not have an atomic command for appending information to the end of a file.

Hard, Soft, and Spongy Mounts

If the NFS client still receives no acknowledgment, it will retransmit the request again and again, each time doubling the time that it waits between retries. If the network filesystem was mounted with the soft option, the request will eventually time out. If the network filesystem is mounted with the hard option, the client continues sending the request until the client is rebooted or gets an acknowledgment. BSDI and OSF/1 also have a spongy option that is similar to hard, except that the stat, lookup, fsstat, readlink, and readdir operations behave like a soft MOUNT.

9.6. NFS PROTOCOL STACK

NFS uses the mount command to specify if a filesystem is mounted with the hard or soft option. To mount a filesystem soft, specify the soft option. For example:

```
/etc/mount -o soft zeus:/big /zbig
```

This command mounts the directory */big* stored on the server called zeus locally in the directory */zbig*. The option -o soft tells the mount program that you wish the filesystem mounted soft.

To mount a filesystem hard, do not specify the soft option:

```
/etc/mount zeus:/big /zbig
```

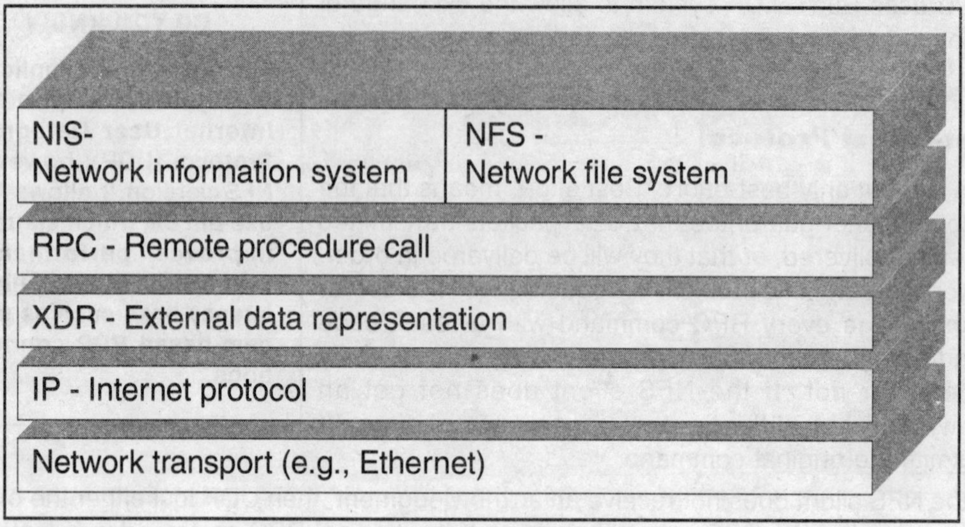

Fig. 9.1.

Deciding whether to mount a filesystem hard or soft can be difficult, because there are advantages and disadvantages to each option. Diskless workstations often hard-mount the directories that they use to keep system programs; if a server crashes, the workstations wait until the server is rebooted, then continue file access with no problem. Filesystems containing home directories are usually hard mounted, so that all disk writes to those filesystems will be correctly performed.

On the other hand, if you mount many filesystems with the hard option, you will discover that your workstation may stop working every time any server crashes until it reboots. If there are many libraries and archives that you keep mounted on your system, but which are not critical, you may wish to mount them soft. You may also wish to specify the intr option, which is like the hard option except that the user can interrupt it by typing the kill character (usually control-C).

As a general rule of thumb, read-only filesystems can be mounted soft without any chance of accidental loss of data. But you will have problems if you try to run programs off partitions that are soft-mounted, because when you get errors, the program that you are running will crash.

9.7. CONNECTIONLESS AND STATELESS

NFS servers are *stateless* by design. Stateless means that all of the information that the client needs to mount a remote filesystem is kept on the client, instead of having additional information with the mount stored on the server. After a file handle is issued for a file, that file handle will remain good even if the server is shut down and rebooted, as long as the file continues to exist and as long as no major changes are made to the configuration of the server that would change the values (e.g., a file system rebuild or restore from tape).

Early NFS servers were also *connectionless*. Connectionless means that the server program does not keep track of every client that has remotely mounted the filesystem. When offering NFS over a TCP connection, however, NFS is not connectionless: there is one TCP connection for each mounted filesystem.

An NFS server computer does keep track of clients that mount their filesystems remotely. The */usr/etc/rpc*.mountd program maintains this database; however, a computer that is not in this database can still access the server's filesystem even if it is not registered in the rpc.mountd database.

The advantage of a stateless, connectionless system is that

1. These systems are easier to write and debug. The programmer does not need to write any code for reestablishing connections after the network server crashes and restarts, because there is no connection that must be reestablished. If a client should crash (or if the network should become disconnected), valuable resources are not tied up on the server maintaining a connection and state for that client.
2. A second advantage of this approach is that it scales. That is, a connectionless, stateless NFS server works equally well if ten clients are using a filesystem or if ten thousand are using it. Although system performance suffers under extremely heavy use, every file request made by a client using NFS will eventually be satisfied, and there is absolutely no performance penalty if a client mounts a filesystem but never uses it.

9.7.1. NFS Servers and Clients

The terms ***client*** and ***server*** are used to describe the roles that a computer plays when sharing file systems. If a file system resides on a computer's disk and that computer makes the file system available to other computers on the network, then that computer acts as a server. The computers that are accessing that file system are said to be clients.

NFS software enables any given computer to access any other computer's file systems and, at the same time, to provide access to its own file systems. A computer may play the role of client, server, or both at any given time on a network.

A server can provide files to a diskless client, a computer that has no local disk. A diskless client relies completely on the server for all its file storage. A diskless client can act only as a client—never as a server. Clients access files on the server by mounting the server's shared file systems. When a client mounts a remote file system, it does not make a copy of the file system; rather, the mounting process uses a series of remote procedure calls that enable the client to access the file system transparently on the server's disk. The mount looks like a local mount and users type commands as if the file systems were local.

Once a file system has been shared on a server through NFS operation, it can be accessed from a client.

> **DO YOU KNOW**
>
> **NFS file systems may be mounted automatically with autofs.**

9.8. AUTOFS

File systems shared through NFS software can be mounted using automatic mounting. Autofs, a client-side service, is a file system structure that provides advanced automatic mounting. The file system structure is created by automount. This program notifies the automount daemon, automountd, when mounting or unmounting needs to be done. The daemon runs in the background, mounting and unmounting remote directories on an as-needed basis.

Whenever a user on a client computer running automountd tries to access a remote file or directory, the daemon mounts the file system to which that file or directory belongs. This remote file system remains mounted for as long as it is needed. If the remote file system is not accessed for a certain period of time, it is automatically unmounted.

No mounting is done at boot time, and the user no longer has to know the superuser password to mount a directory; users need not use the mount and umount commands. The autofs service mounts and unmounts file systems as required without any intervention on the part of the user.

Mounting some file hierarchies with automountd does not exclude the possibility of mounting others with mount. A diskless computer ***must*** mount / (root), /usr, and /usr/kvm through the mount command and the /etc/vfstab file

9.8.1. Autofs Features

Autofs works with file systems specified in the local name space. This information can be maintained in NIS, NIS+, or local files.

The name space data can specify several remote locations for a particular file. This way, if one of the servers is down, the autofs service can try to mount from another computer. To specify which servers are preferred for each file system in the maps, each server can be assigned a weighting factor

9.9. WORKING OF NFS

NFS consists of at least two main parts:

- A server and one or more
- Clients.

The client remotely accesses the data that is stored on the server machine. In order for this to function properly a few processes have to be configured and running:

The server has to be running the following commands:

Commands	Description
nfsd	The NFS coomand which services requests from the NFS clients.
mountd	The NFS mount command which carries out the requests that nfsd passes on to it.
portmap	The portmapper daemon allows NFS clients to discover which port the NFS server is using.

The client can also run a command, known as **nfsiod**. The **nfsiod** command services the requests from the NFS server. This is optional, and improves performance, but is not required for normal and correct operation.

How Does NFS Work ?

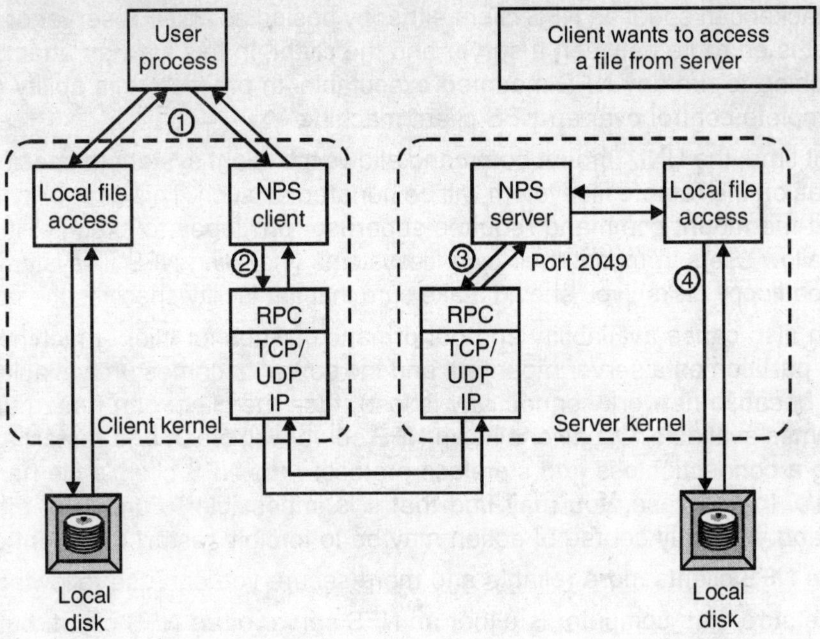

Fig. 9.2.

9.10. SERVER-SIDE NFS SECURITY [MDU Dec. 2006]

Because NFS allows users on a network to access files stored on the server, NFS has significant security implications for the server. These implications fall into three broad categories:

1. Client Access

NFS can (and should) be configured so that only certain clients on the network can mount filesystems stored on the server.

2. User Authentication [MDU Dec. 2006]

NFS can (and should) be configured so that users can only access and alter files to which they have been granted access.

3. Eavesdropping and Data Spoofing

NFS should (but does not) protect information on the network from eavesdropping and surreptitious modification.

9.11. CLIENT-SIDE NFS SECURITY

NFS can create security issues for NFS clients as well as servers. Because the files that a client mounts appear in the client's filesystem, an attacker who is able to modify mounted files can directly compromise the client's security.

The primary system that NFS uses for authenticating servers is based on IP host addresses and hostnames. NFS packets are not encrypted or digitally signed in any way. Thus, an attacker can spoof an NFS client either by posing as an NFS server or by changing the data that is en route between a server and the client. In this way, an attacker can force a client machine to run any NFS-mounted executable. In practice, this ability can give the attacker complete control over an NFS client machine.

At mount time, the UNIX mount command allows the client system to specify whether or not SUID files on the remote filesystem will be honored as such. This capability is one of the reasons that the mount command requires superuser privileges to execute. If you provide facilities to allow users to mount their own filesystems (including NFS filesystems as well as filesystems on floppy disks), you should make sure that the facility specifies the nosuid option.

NFS can also cause availability and performance issues for client machines. If a client has an NFS partition on a server mounted, and the server becomes unavailable (because it crashed, or because network connectivity is lost), then the client can freeze until the NFS server becomes available. Occasionally, an NFS server will crash and restart and - despite NFS's being a connectionless and stateless protocol - the NFS client's file handles will all become *stale*. In this case, you may find that it is impossible to unmount the stale NFS filesystem, and your only course of action may be to forcibly restart the client computer.

To make NFS clients more reliable and more secure you can use following options:

- Make sure that computer is either an NFS server or an NFS client, but not both.
- If possible, do not allow users to log into your NFS server.
- Don't allow NFS clients to mount NFS servers outside your organization.
- Minimize the number of NFS servers that each client mounts. A system is usually far more reliable and more secure if it mounts two hard disks from a single NFS server, rather than mounting partitions from two NFS servers.
- If possible, disable the honoring of SUID files and devices on mounted partitions.

9.12. IMPROVING NFS SECURITY

There are many techniques that you can use to improve overall NFS security:

1. Limit the use of NFS by limiting the machines to which filesystems are exported, and limit the number of filesystems that each client mounts.
2. Export filesystems read-only if possible.
3. Use root ownership of exported files and directories.
4. Remove group write permissions from exported files and directories.
5. Do not export the server's executables.
6. Do not export home directories.
7. Do not allow users to log into the server.
8. Use the fsirand program, .
9. Set the portmon variable, so that NFS requests that are not received from privileged ports will be ignored.
10. Use showmount -e to verify that you are only exporting the filesystem you wish to export to the hosts specified with the correct flags.
11. Use Secure NFS.

1. Limit Exported and Mounted Filesystems

The best way to limit the danger of NFS is by having each computer only export and/or mount the particular filesystems that are needed.

If a filesystem does not need to be exported, do not export it. If it must be exported, export it to as few machines as possible by judiciously using restrictions in the exports list. If you have a sizeable number of machines to export to and such lists are tedious to maintain, consider careful use of the netgroups mechanism, if you have it. Do not export a filesystem to any computer unless you have to. If possible, export filesystems read-only.

If you only need to export part of a filesystem, then export only that part. Do not export an entire filesystem if you only need access to a particular directory.

Likewise, your clients should only mount the NFS servers that are needed. Don't simply have every client in your organization mount every NFS server. Limiting the number of mounted NFS filesystems will improve overall security, and will improve performance and reliability as well.

2. Export Read-only

Many filesystems contain information that is only read, never (or rarely) written. These filesystems can be exported read-only. Exporting the filesystems read-only adds to both security and reliability: it prevents the filesystems from being modified by NFS clients, limiting the damage that can be done by attackers, ignorant users, and buggy software.

Many kinds of filesystems are candidates for read-only export:

- Filesystems containing applications
- Organizational reference matter, such as policies and documents
- Netnews (if you do not read news with NNTP)

3. Use Root Ownership

Because the NFS server maps *root* to *nobody*, you can protect files and directories on your server by setting their owner to root and their protection mode to 755 (in the case of programs and directories) or 644 (in the case of data files). This setup will prevent the contents of the files from being modified by a client machine.

If you have information on an NFS server that should not be accessible to NFS clients, you can use the file protection mode 700 (in the case of programs and directories) or 600 (in the case of data files). However, a better strategy is not to place the files on the NFS server in the first place.

4. Remove Group-Write Permission for Files and Directories

If you are using standard AUTH_UNIX authentication with NFS, then users can effectively place themselves in any group. Thus, to protect files and directories that are owned by *root*, they must *not* be group-writable.

5. Do Not Export Server Executables

If our server is running the same operating system on the same CPU architecture as your client computers, then you might be tempted to have the server export its own executables (such as the programs stored in */bin, /usr/bin, /etc.*) for use by the clients. Don't do so without careful thought about the consequences.

But exporting your server's executables poses several security problems:

- It allows an attacker to easily determine which version of each executable your server is running, which enables the attacker to probe for weak spots with greater ease.
- If there is an error in your system's configuration, you may be exporting the binaries on a writable filesystem. An attacker could then modify the server's own binaries, and possibly break in .

6. Do Not Export Home Directories

If you export a filesystem that has users' home directories on it and you do not use Secure RPC, then all other clients mounting that directory, as well as the server itself, can be placed at risk.

If you export a filesystem that contains users' home directories, then there is a risk that an attacker could alter the information stored on the NFS server. This is normally a serious risk in itself. However, if the partition being exported includes users' home directories, then one of the things that an attacker can do is create files in the users' home directories.

A simple attack is for an attacker to create a *.rhosts* file in a users's home directory that specifically allows access to the attacker. Having created this file, the attacker could now log onto the server and proceed to look for additional security holes. Perhaps the greatest danger in this attack is that it can be aimed against system accounts (such as *daemon* and *bin*) as easily as accounts used by human users.

7. Do Not Allow Users to Log into the Server

NFS and direct logins are two fundamentally different ways to use a computer. If you allow users to log into a server, the user can use that access to probe for weaknesses that can be exploited from NFS, and vice versa.

8. Use fsirand

One of the security problems with NFS is that the file handles used to reference a file consist solely of a filesystem ID, an inode number, and a generation count. Guessing valid file handles is easy in most circumstances. Filesystem IDs are normally small numbers; the root directory on the standard UNIX filesystem has the inode number 2, */lost+found* has the inode number 3, and so on.

The fsirand program increases the difficulty of guessing a valid file handle by randomizing the generation number of every inode on a filesystem. The effect is transparent to the user - files and directories are still fetched as appropriate when a reference is made - but someone on the outside is unable to guess file handles for files and directories anymore.

You can run fsirand on the *root* directory while in single-user mode or on any unmounted filesystem that will fsck without error.

9. Set the portmon Variable

Normally, NFS servers respond to requests that are transmitted from any UDP port. However, because NFS requests are supposed to come from *kernels* of other computers, and not from *users* who are running user-level programs on other computers, a simple way to improve the security of NFS servers is to program them to reject NFS requests that do not come from privileged ports. On many NFS servers, the way that this restriction is established is by setting the kernel variable *nfs_portmon* to 1.

10. Use Showmount -e

The *showmount -e* command, , lists the host"s export lists - -that is, the directories and hosts that can be mounted. The *showmount* command allows an optional argument, *host*. When this argument is provided, the *showmount* command can be used to remotely inspect another computer's export list. The command is useful for finding NFS servers which are insecurely configured

11. Use Secure NFS

The biggest security problem with NFS, as it is normally configured, is that it uses Sun's AUTH_UNIX RPC authentication system. With AUTH_UNIX, a user simply provides his UID and a list of GIDS with every request. The NFS server trusts that the users are who they claim to be.

In a friendly environment, AUTH_UNIX authentication presents no problems, because requests sent out by the NFS client always have the same UID and GIDS as the person who has logged in and is using the workstation. However, if the workstation user has *root* access, that person can use the root access to become any other user, with that other user's corresponding rights and privileges on the RPC server. A second problem with AUTH_UNIX is that user-written programs can have their AUTH_UNIX UID and GIDS set to any value. When reserved port checking is enabled, AUTH_UNIX offers roughly the same level of security as the *rsh/rlogin* trusted-host facility.

[Secure NFS overcomes these problems by using AUTH_DES RPC authentication instead of AUTH_UNIX. With Secure NFS, users must be able to decrypt a special key stored on the NIS or NIS+ server before the NFS filesystem will allow the user to access his or her.

NFS depends on NIS or NIS+ on many machines. Both NFS and NIS implementations have had some well-known implementation flaws and bugs in recent years. Not only are these flaws well-known, there are a number of hacker toolboxes available that include programs to take advantage of these flaws. Therefore, if you are running NFS, it should be certain that it is up to date on vendor patches and bug fixes. In particular:

- Make sure that version of the RPC portmapper does not allow *proxy requests* and that your own system is not in the export list for a partition. Otherwise, a faked packet sent to your RPC system can be made to fool your NFS system into acting as if the packet was valid and came from your own machine.
- Make sure that your NFS either uses Secure RPC, or examines the full 32 bits of the UIDS passed in. Some early versions of NFS only examined the least significant 16 bits of the passed-in UID for some tests, so accesses could be crafted that would not get mapped to *nobody*, and would function as root accesses.
- Make sure that version of NFS does not allow remote users to issue mknod commands on partitions they import from your servers. A user creating a new */dev/kmem* file on partition has made a big first step towards a complete compromise of your system.

➢ Make sure that NFS does the correct thing when someone does a cd.. in the top level of an imported directory from your server. Some older versions of NFS would return a file handle to the server's real parent directory instead of the parent to the client's mount point. Because NFS doesn't know how you get file handles, and it applies permissions on whole partitions rather than mount points, this process could lead to your server's security being compromised.

In particular, when a server would export a subdirectory as the *root* partition for a diskless workstation, a user on the workstation could do cd /; cd... and instead of getting the *root* directory again, they would have access to the parent directory on the server! Further compounding this scenario, the export of the partition needed to be done with *root=* access. As a result, clients would have unrestricted access to the server's disks!

➢ Make sure that server parses the export option list correctly. Some past and current NFS implementations get accesses mixed up. If you specify *access=* and *rw=* on the same export, or *access=* and *root=,* the system sometimes forgets the *access=* specification and exports the partition to every other machine in the world.

9.13. THE SOCKET API

The socket layer provides an API (application program interface) for user programs to access the networking functionality of the kernel. for example, socket(), bind(), accept(), etc. calls are "system calls" provided by the kernel so that user programs can use kernel functionalities.

For user programs to interface to the simulator, you can use the following socket calls: Socket(),

```
Close(),
Bind(),
Read(),
Write(),
Sendto(),
Recvfrom(),
and Setsockopt().
```

Their prototypes are defined in $PDIR/include/Socket.h (this header file should be included by user programs). There are two important issues regarding these calls:

➢ Observe that the first letter of each call is *capitalized.* This is to distinguish them from the actual Solaris system calls, which will go into the Solaris kernel upon invocation.

The Socket() Call

The Socket() call accepts three arguments: *family*, *type*, and *protocol.* It supports the following three combinations of *family* and *type*:

1. **AF_INET/SOCK_STREAM :** This combination specifies that the user wants to create a TCP socket. The following system calls are allowed on a TCP socket: Close(), Bind(), Accept(), Connect(), Write(), Read(), and Setsockopt().

2. **AF_INET/SOCK_DGRAM :** This combination specifies that the user wants to create a UDP socket. The following system calls are allowed on a UDP socket: Close(), Bind(), Sendto(), Recvfrom(), and Setsockopt().
3. **AF_INET/SOCK_ICMP :** This combination specifies that the user wants to create an ICMP socket. The following system calls are allowed on an ICMP socket: Close() and Recvfrom().
4. **AF_ROUTE/SOCK_RAW :** This combination specifies that the user wants to create a routing socket. The following system calls are allowed on a routing socket: Close(), Write(), and Setsockopt().

The Recvfrom() Call

Here is the prototype of the Recvfrom() system call:

```
int Recvfrom(int s, void *buf, int len, int flags, struct
sockaddr                                            *from,
int *fromlen);
```

The arguments to this call are basically the same as the standard socket call. The Recvfrom() call reads "one packet at a time". It returns the length of the message written to the buffer pointed to by the buf argument (the second argument). Even if one packet worth of message does not "fill up" the buffer, Recvfrom() will return immediately and will not read the second packet. However, if a message in a packet is too long to fit in the supplied buffer, the excess bytes are discarded.

By default, Recvfrom() is blocking: when a process issues a Recvfrom() that cannot be completed immediately (because there is no packet), the process is put to sleep waiting for a packet to arrive at the socket. Therefore, a call to Recvfrom() will return immediately *only if* a packet is available on the socket.

When the argument flags of Recvfrom() is set to MSG_NOBLOCK, Recvfrom() does *not* block if there is no data to be read, but returns immediately with a return value of 0 bytes. MSG_NOBLOCK is defined in $PDIR/include/systm.h. In an actual UNIX system, socket descriptors are set to be non-blocking using fcntl() with type O_NONBLOCK, and Recvfrom() returns errno EWOULDBLOCK when there is no data to be read on the non-blocking socket.

The Sendto() Call

The Sendto() call has an argument flags, which is ignored by the current implementation.

The Write() call

The Write() call cannot be used with a UDP socket. This is quite different from the standard write() call in UNIX, which can be used with any socket.

TIPS

The Write() call can only be used with a routing or a TCP socket.

The Setsockopt() call

The Setsockopt() call has five arguments. The arguments level and optname specify the option that you want to set. The simulator supports only the IPPROTO_IP level and the IP_FW_SET, IP_NAT_SET, and IP_IF_SET options (defined in $PDIR/include/Socket.h). The option value for the IP_IF_SET option needs to be a pointer to a struct if_info (defined in $PDIR/include/route.h). The data structures of the option values for IP_FW_SET and IP_NAT_SET have not been defined.

9.13.1. TCP

There are two types of applications that use TCP sockets :

- Servers and
- Clients.

> **DO YOU KNOW**
>
> **A server socket in the simulator accepts only one TCP connection in its lifetime.**

A TCP server listens on a well-known port (or IP address and port pair) and accepts connections from TCP clients. A TCP client initiates a connection request to a TCP server in order to setup a connection with the server. A real TCP server can accept multiple connections on a socket.

Below is the sequence of socket calls made by a TCP server and a TCP client:

```
Server: Socket -> Bind -> Accept -> Read/Write -> Close
Client: Socket -> (Bind) -> Connect -> Read/Write -> Close
```

Below are the details of the socket calls specification.

A server must call Bind() in order to bind the socket to a port, before calling Accept(). Otherwise, Accept() returns an error. Bind() to a client socket is optional.

We make the following changes to the Accept() specification :

1. Accept() returns 0 on success (instead of a new file descriptor), and -1 on failure. Accept() does not create a new file descriptor (unlike the Berkeley Socket specification), and uses the same file descriptor for all subsequent socket calls.
2. The socket starts accepting client connection requests only *after* the Accept() call has been made (i.e. packets arriving before the Accept() call are discarded).
3. If during the connection establishment phase the protocol times out (e.g. during the TCP three-way handshake) or an error occurs, Accept() waits for other connection attempts, and returns only when a connection has been established successfully.
4. Accept() is always blocking—Accept() should block until after a connection establishment is completed.

Connect() is always blocking. It returns 0 if the connection establishment (TCP three-way handshake) succeeds, and returns -1 if an error occurs. One possible error is a timeout during the three-way handshake. In all cases where errors occur during the Connect() call, an application should not call Connect() again but should call Close() immediately.

Read() is always blocking. On success, the number of bytes read is returned. If Read() returns 0, this indicates an "End of File", meaning that the other side has closed the connection, and no more data will be received by this socket.

Write() returns almost immediately, except when the *send buffer* is full. This buffer is used by TCP to keep all data that could not be sent immediately (because of window limitations), as well as, data that has been sent but has not yet been acknowledged. If the send buffer is full, Write() blocks until the send buffer is freed to enqueue another packet.

When a user program calls Close() on a TCP socket, the socket is marked as closed and the Close() function returns to the process immediately. The socket descriptor is no longer usable by the process, i.e. it would not be usable as an argument to Read() or Write(). TCP tries to send all data that is already queued to be sent to the other end, and after this occurs the normal TCP connection termination sequence takes place

9.14. ROUTING SOCKETS [MDU Dec. 2007; May 2008; Dec. 2008; May 2006]

Routing sockets are created with domain

- AF_ROUTE and
- type SOCK_RAW.

This type of socket is used with the following system calls:

- Socket(),
- Close(),
- Write(), and
- Setsockopt().

A routing socket is a special type of socket that is not specific to any particular network protocol, but allows "privileged" user processes to write information into the kernel. User processes use this type of socket to add and remove information from the routing table. This is done by filling in the rt_msghdr structure and passing it to Write(). The rt_msghdr structure is defined in $PDIR/include/route.h.

The following values of the rtm_type field of the rt_msghdr structure are supported:

- RTM_ADD: add an entry to the routing table;
- RTM_DELETE: delete an entry from the routing table;
- RTM_CHANGE: change an entry in the routing table. The implementation of this command is equivalent to performing an RTM_DELETE followed by an RTM_ADD.

These constants are defined in $PDIR/include/route.h.

To look up a route within the kernel, call rt_lookup_dest() (defined in $PDIR/include/rtable.h). This function will return the address of the gateway to which a packet with the given destination address should be sent.It also returns the index of the interface over which the packet should be sent.

Three types of operations are possible on routing sockets.

1. Process can send a message to the kernel by writing to a routing socket.

DO YOU KNOW

The current implementation of the simulator supports only host routes, there is no support for network masks or prefix matching. If there is no host route available in the routing table, the function returns the default gateway. The default gateway route is defined as a route that has a destination address of 0.0.0.0. It can be inserted into the routing table in the same way as host routes.

2. Process can read a message from the kernel on a routing socket.
3. Process can use the system 1 function to either dump the routing table or to list all the configured interface.

> **DO YOU KNOW**
>
> **After a process create a routing socket, it can send comands to the kernal by writing to the socket and read information from the kernel by reading from the socket.**

ICMP

ICMP is an integral part of any IP implementation. ICMP is normally used to communicate error messages between IP nodes routers and endhosts, but it is occasionally used by user-level applications such as traceroute.

An ICMP message can be either a

- *query* message or an
- *error* message,

and it has two fields

- *type* field
- *code* field.

To send an ICMP message within the kernel, call icmp_send() and pass it the packet causing the error condition and the desired type and code.

There is also a mechanism for a user-level process to read ICMP packets received by a host. In a real UNIX socket implementation, a process opens a RAW socket to receive ICMP packets. In the simulator, there is a new socket type SOCK_ICMP, which is defined in $PDIR/include/systm.h. You can open an ICMP socket as follows:

```
Socket(AF_INET, SOCK_ICMP, 0);
```

The ICMP socket implementation also supports the Recvfrom() and Close() socket calls. Note that the ICMP header of a received packet is not stripped by the kernel so that a user-level process can access the ICMP header to see the error type and code, among other things.

If an ICMP packet arrives at a host, and no ICMP socket is opened, the host drops the packet.

> **DO YOU KNOW**
>
> **There is no need to bind an ICMP socket to a particular IP address: the ICMP socket will get ICMP messages received by any of the host's interface(s). As a result, you should make sure that at most one ICMP socket is opened at any given time.**

9.15. INTRODUCTION TO MOBILE NETWORK

Mobile means able to move freely or easily. There are two aspect of mobility:

- *User mobility*: users communicate (wireless) "anytime, anywhere, with anyone"
- *Device portability*: devices can be connected anytime, anywhere to the network

Wireless vs.	mobile	Examples
×	×	stationary computer
×	✓	notebook in a hotel
✓	×	wireless LANs in historic buildings
✓	✓	Personal Digital Assistant (PDA)

The demand for mobile communication creates the need for integration of wireless networks into existing fixed networks:

➢ **Local area networks :** Standardization of IEEE 802.11, ETSI (HIPERLAN)

➢ **Internet :** Mobile IP extension of the internet protocol IP

➢ **Wide area networks :** For example, internetworking of GSM and ISDN

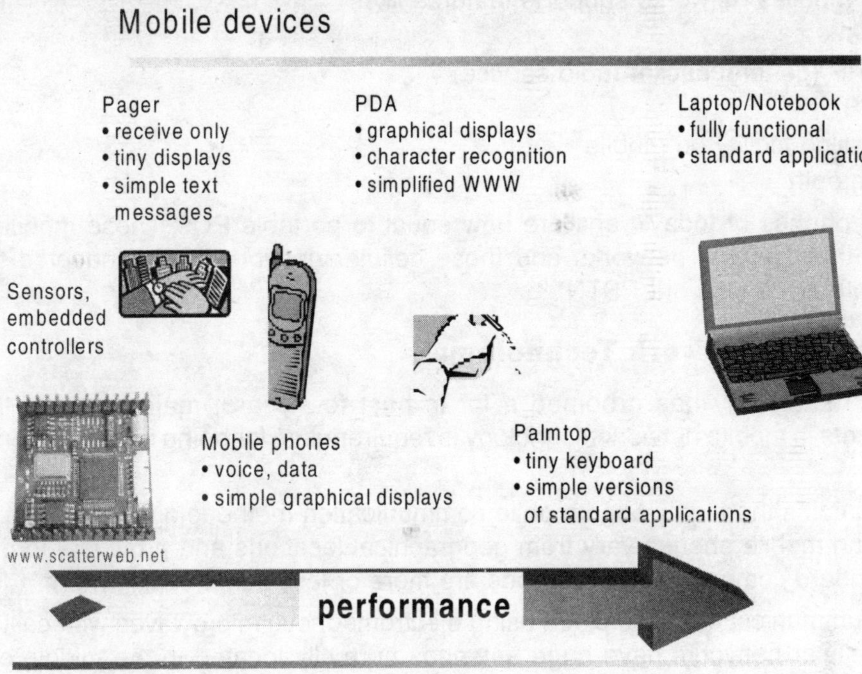

Fig. 9.3.

Comparison of Wireless Network with Fixed Network

Higher loss-rates in wireless due to interference

➢ emissions of, e.g., engines, lightning

Restrictive regulations of frequencies in wireless network

➢ frequencies have to be coordinated, useful frequencies are almost all occupied

Low transmission rates in wireless network

➢ local some Mbit/s, regional currently, e.g., 53kbit/s with GSM/GPRS

Higher delays, higher jitter in wireless network

➢ connection setup time with GSM in the second range, several hundred milliseconds for other wireless systems

Lower security, simpler active attacking

➢ radio interface accessible for everyone, base station can be simulated, thus attracting calls from mobile phones

Always shared medium

➢ secure access mechanisms important

Mobile technology has already creating hype through out the world. We are using so many features in our mobile these days which most of us have never dreamed off. We are using SMS, deliver messages in seconds through out the world with just a little click of button"Send", MMS, sending pictures and video files in amazingly quick time, finding locations, accessing high speed internet in your mobiles are the features which were not possible just few years back.

Today's mobiles networks supports features likes

- SMS
- GPRS (general packet radio service)
- MMS
- Emailing facility on mobile
- Bluetooth

Mobile phones of today's age are now equal to portable PCs. These mobile phones connect to their cellular networks and these cellular networks are connected to Public switching telephone network PSTN).

9.15.1. Mobile Network Technology

Mobile Technology has groomed a lot in past few years, major reasons for rapid advancements in mobile network technology is requirements for being mobile or connectivity on move.

GSM remains the highly used mobile communication methodology worldwide. Cellular networks and mobile phones vary from geographical locations and providers to providers, but still standard communication methods are more or less same every where.

Basic communication takes place using electromagnetic microwaves with cellular base stations. Cellular networks have huge antennas normally located in the middle of certain area to provide optimum signal broadcasting. These antennas are known as Base Transceiver Station (BTS). Mobile handsets have low powered transceivers which transmit voice data to the closed BTS which can usually be with in 5 t 8 miles radius.

Transceivers: Transceivers are devices which have capability of transmitting and receive data at the same time.Mobile handset register it self at mobile network switch or exchange as soon as it starts transmitting, usually it means when handset is turn on, it tries to registered to the network of inserted SIM card. Mobile networks uniquely identify each and every registered mobile handset on the given BTS and sends notifications to those mobile handsets which have incoming calls. Handsets will always be registered to the most nearest BTS for communication, incase if user of the network is on move, mobile phones in this case will handoff to many closest BTS during the call and keep the connection alive with nearest base stations. In case of unavailability of BTS, handset will be disconnected from the network resulting in dropping of on going call.

Communication between BTS and mobile sets takes place in stream of digital data, which is digitized audio. Each mobile network operates on unique radio frequency.

Broadcasting in Mobile Network

There are two information delivery methods for wireless data applications:

- point-to-point access
- broadcast.

In point-to-point access, a logical channel is established between the client and the server. Queries are submitted to the server and results are returned to the client in much the same way as in a wired network.

In broadcast, data are sent simultaneously to all users residing in the broadcast area. It is up to the client to select the data it wants. the client can also submit queries to the server so that the data it wants are guaranteed to be broadcast.

Compared with point-to-point access, broadcast is a better method for several reasons:

1. A single broadcast of a data item can satisfy all the outstanding requests for that item simultaneously. As such, broadcast can scale up to an arbitrary number of users.
2. Mobile wireless environments are characterized by asymmetric communication, i.e., the downlink communication capacity is much greater than the uplink communication capacity. Data broadcast can take advantage of the large downlink capacity when delivering data to clients.
3. A wireless communication system essentially employs a broadcast component to deliver information. Thus, data broadcast can be implemented without introducing any additional cost.

Although point-to-point and broadcast systems share many concerns, such as the need to improve response time while conserving power and bandwidth consumption.

Access efficiency and power conservation are two critical issues in any wireless data system. Access efficiency concerns how fast a request is satisfied, and power conservation concerns how to reduce a mobile client's power consumption when it is accessing the data it wants. The second issue is important because of the limited battery power on mobile clients, which ranges from only a few hours to about half a day under continuous use.

Two Performance Metrics

- ➢ Access time
- ➢ tune-in time,

are used to measure access efficiency and power conservation for a broadcast system, respectively:

Access time is the time elapsed between the moment when a query is issued and the moment when it is satisfied.

Tune-in time is the time a mobile client stays active to receive the requested data items.

There are three kinds of broadcast models

- ➢ push-based broadcast
- ➢ on-demand broadcast
- ➢ hybrid broadcast.

In push-based broadcast the server disseminates information using a periodic/aperiodic broadcast program in on-demand broadcast the server disseminates information based on the outstanding requests submitted by clients; in hybrid broadcast push-based broadcast and on-demand data deliveries are combined to complement each other.

HIGHLIGHTS

1. NFS is a technology for sharing resources between devices on a local area network.
2. Advantages of NFS :
   - Data can be kept on central host
   - Administrative data may be kept on single host.
3. Disadvantages of NFS :
   - Right for unwanted people to access hard drive.
   - Security
4. Various versions of NFS
   - NFS V1
   - NFS V2
   - NFS V3
   - NFS V4
5. NFS V4 handles 64 bits files.
6. NFS servers are stateless by design.
7. NFS consits of two parts
   - A server
   - A client
8. There are various methods by which NFS security can be improved.
9. Routing sockets are created with
   - AF-route
   - Type sock-raw
10. Following sysetm calls are used in routing socket
    - socket ( )
    - close ( )
    - write ( )
    - setsockopt ( )
11. Routing socket allows privileged user processes to write information into the kernel.

QUESTIONS WITH ANSWERS

Q. 1. What is the difference between the different NFS versions?

Ans. Currently the NFS version 2 and NFS version 3 protocols are in use with version 4 under consideration for a standard. There are quite a few differences between version 2 and 3 and some of the more notable ones are:

(i) Version 2 uses UDP (User Datagram Protocol) in most implementations which does not have guaranteed delivery.

(ii) Version 3 uses TCP (Transmission Control Protocol) that is a software layer to guarantee delivery.

(iii) Version 3 supports use of asynchronous writes to improve performance.

(iv) Version 2 supports 64-bit file sizes (vs. 32 for v2) to allow much larger files.

For NFS version 4, there are many proposed changes. The NFS version 4 protocol is a further revision of the NFS protocol defined already by versions 2 and 3 . It retains the essential characteristics of previous versions: design for easy recovery, independent of transport protocols, operating systems and file systems, simplicity and good performance.

The NFS version 4 revision has the following goals:

(i) **Improve access and good performance on the Internet.** The protocol is designed to transit firewalls easily, perform well where latency is high and bandwidth is low and scale to very large numbers of clients per server.

(ii) **Strong security with negotiation built into the protocol.** The protocol builds on the work of the ONCRPC working group in supporting the RPCSEC_GSS protocol. Additionally, the NFS version 4 protocol provides a mechanism to allow clients and servers the ability to negotiate security and require clients and servers to support a minimal set of security schemes.

(iii) **Good cross-platform interoperability.** The protocol features a file system model that provides a useful, common set of features that does not unduly favor one file system or operating system over another.

(iv) **Designed for protocol extensions.** The protocol is designed to accept standard extensions that do not compromise backward compatibility.

REVIEW QUESTIONS

1. What is NFS. Explain NFS environment.
2. What are the advantages and disadvantages of NFS.
3. Explain various versions of NFS.
4. Explain various versions of NFS which are available.
5. Difference between NFSv2 and NFSv3 and NFSv4.
6. What is mount protocol.
7. Explain the working NFS client and server.
8. How we can improve NFS security.
9. What is routing socket.
10. Explain server side NFS security.

❑❑❑

APPENDICES

A. FUNDAMENTALS CONCEPTS OF NETWORKING

B. VIRUSES

C. REMOTE PROCEDURE CALL TOOLS

Appendix A
Fundamentals Concepts of Networking

The Network Information Service (NIS) and Network File System (NFS) are services that allow you to build distributed computing systems that are both consistent in their appearance and transparent in the way files and data are shared.

NIS provides a distributed database system for common configuration files. NIS servers manage copies of the database files, and NIS clients request information from the servers instead of using their own, local copies of these files. For example, the */etc/hosts* file is managed by NIS. A few NIS servers manage copies of the information in the hosts file, and all NIS clients ask these servers for host address information instead of looking in their own */etc/hosts* file. Once NIS is running, it is no longer necessary to manage every */etc/hosts* file on every machine in the network — simply updating the NIS servers ensures that all machines will be able to retrieve the new configuraton file information.

NFS is a distributed filesystem. An NFS server has one or more filesystems that are mounted by NFS clients; to the NFS clients, the remote disks look like local disks. NFS filesystems are mounted using the standard Unix *mount* command, and all Unix utilities work just as well with NFS-mounted files as they do with files on local disks. NFS makes system administration easier because it eliminates the need to maintain multiple copies of files on several machines: all NFS clients share a single copy of the file on the NFS server. NFS also makes life easier for users: instead of logging on to many different systems and moving files from one system to another, a user can stay on one system and access all the files that he or she needs within one consistent file tree.

NFS is core of a distributed computing environment, because it manages the resource users are most concerned about: their files. However, a distributed filesystem such as NFS will not function properly if hosts cannot agree on configuration information such as usernames and host addresses. The primary function of NIS is managing configuration information and making it consistent on all machines in the network. NIS provides the framework in which to use NFS.

Networking Overview

NFS and NIS are high-level networking protocols, built on several lower-level protocols. In order to understand the way the high-level protocols function, you need to know how the underlying services work. The lower-level network protocols are quite complex, and several books have been written about them without even touching on NFS and NIS services. Therefore, this chapter contains only a brief outline of the network services used by NFS and NIS.

Network protocols are typically described in terms of a layered model, in which the protocols are "stacked" on top of each other. Data coming into a machine is passed from the lowest-level protocol up to the highest, and data sent to other hosts moves down the protocol stack. The layered model is a useful description because it allows network services to be defined in terms of their functions, rather than their specific implementations. New protocols can be substituted at lower levels without affecting the higher-level protocols, as long as these new protocols behave in the same manner as those that were replaced.

The standard model for networking protocols and distributed applications is the International Organization for Standardization (ISO) seven-layer model shown in Table below

TABLE 1. The ISO Seven-Layer Model

Layer	Name	Physical Layer
7	Application	NFS and NIS
6	Presentation	XDR
5	Session	RPC
4	Transport	TCP or UDP
3	Network	IP
2	Data Link	Ethernet
1	Physical	CAT-5

The lower levels have a well-defined job to do, and the higher levels rely on them to perform it independently of the particular medium or implementation. While TCP/IP most frequently is run over Ethernet, it can also be used with a synchronous serial line or fiber optic network. Different implementations of the first two network layers are used, but the higher-level protocols are unchanged. Consider an NFS server that uses all six lower protocol layers: it has no knowledge of the physical cabling connecting it to its clients. The server just worries about its NFS protocols and counts on the lower layers to do their job as well.

Understanding the basic structure of the network services on which NFS and NIS are built is essential for designing and configuring large networks, as well as debugging problems. A failure or overly tight constraint in a lower-level protocol affects the operation of all protocols above it. If the physical network cannot handle the load placed on it by all of the desktop workstations and servers, then NFS and NIS will not function properly. Even though NFS or NIS will appear "broken," the real issue is with a lower level in the network stack.

Physical and Data Link Layers

The physical and data link layers of the network protocol stack together define a machine's *network interface*.

From a software perspective, the network interface defines how the Ethernet device driver gets packets from or to the network. The physical layer describes the way data is actually transmitted on the network medium. The data link layer defines how these streams of bits are put together into manageable chunks of data.

Ethernet is the best known implementation of the physical and data link layers. The Ethernet specification describes how bits are encoded on the cable and also how stations on the network detect the beginning and end of a transmission.

Ethernet can be run over a variety of media, including

- thinnet,
- thicknet,
- unshielded twisted-pair (UTP) cables, and
- fiber optics.

All Ethernet media are functionally equivalent — they differ only in terms of their convenience, cost of installation, and maintenance. Converters from one media to another operate at the physical layer, making a clean electrical connection between two different kinds of cable. Unless you have access to high-speed test equipment, the physical and data link layers are not that interesting when they are functioning normally. However, failures in them can have strange, intermittent effects on NFS and NIS operation.

Frames and Network Interfaces

The data link layer defines the format of data on the network. A series of bits, with a definite beginning and end, constitutes a network *frame*, commonly called a *packet*. A proper data link layer packet has checksum and network-specific addressing information in it so that each host on the network can recognize it as a valid (or invalid) frame and determine if the packet is addressed to it. The largest packet that can be sent through the data link layer defines the *Maximum Transmission Unit*, or MTU, of the network.

All hosts have at least one network interface, although any host connected to an Ethernet has at least two: the *Ethernet interface* and the *loopback interface*. The Ethernet interface handles the physical and logical connection to the outside world, while the loopback interface allows a host to send packets to itself. If a packet's destination is the local host, the data link layer chooses to "send" it via the loopback, rather than Ethernet, interface. The loopback device simply turns the packet around and queues it at the bottom of the protocol stack as if it were just received from the Ethernet.

You may find it helpful to think of the protocol layers as passing packets upstream and downstream in envelopes, where the packet envelope contains some protocol-specific header information but hides the remainder of the packet contents. As data messages are passed from the top most protocol layer down to the physical layer, the messages are put into envelopes of increasing size. Each layer takes the entire message and envelope from the layer above and adds its own information, creating a new message that is slightly larger than the original. When a packet is received, the data link layer strips off its envelope and passes the result up to the network layer, which similarly removes its header information from the packet and passes it up the stack again.

Ethernet Addresses

Associated with the data link layer is a method for addressing hosts on the network. Every machine on an Ethernet has a unique, 48-bit address called its *Ethernet* or *Media Access Control (MAC) address*. Vendors making network-ready equipment ensure that every machine in the world has a unique MAC address. 24-bit prefixes for MAC addresses are assigned to hardware vendors, and each vendor is responsible for the uniqueness of the lower 24 bits. MAC addresses are usually represented as colon-separated pairs of hex digits:

```
8:0:20:ae:6:1f
```

Note that MAC addresses identify a *host*, and a host with multiple network interfaces may use the same MAC address on each.

Part of the data link layer's protocol-specific header are the packet's source and destination MAC addresses. Each protocol layer supports the notion of a *broadcast*, which is a packet or set of packets that must be sent to all hosts on the network. The broadcast MAC address is:

```
ff:ff:ff:ff:ff:ff
```

All network interfaces recognize this wildcard MAC address as a broadcast address, and pass the packet up to a higher-level protocol handler.

Network Layer

At the data link layer, things are fairly simple. Machines agree on the format of packets and a standard 48-bit host addressing scheme. However, the packet format and encoding vary with different physical layers: Ethernet has one set of characteristics, while an X.25-based satellite network has another. Because there are many physical networks, there should ideally be a standard interface scheme so that it isn't necessary to re-implement protocols on top of each physical network and its peculiar interfaces. This is where the network layer fits in. The higher-level protocols, such as TCP (at the transport layer), don't need to know any details about the physical network that is in use. As mentioned before, TCP runs over Ethernet, fiber optic network, or other media; the TCP protocols don't care about the physical connection because it is represented by a well-defined network layer interface.

The network layer protocol of primary interest to NFS and NIS is the Internet Protocol, or IP. As its name implies, IP is responsible for getting packets between hosts on one or more networks. Its job is to make a best effort to get the data from point A to point B. IP makes no guarantees about getting all of the data to the destination, or the order in which the data arrives — these details are left for higher-level protocols to worry about.

On a local area network, IP has a fairly simple job, since it just moves packets from a higher-level protocol down to the data link layer. In a set of connected networks, however,

IP is responsible for determining how to get data from its source to the correct destination network. The process of directing datagrams to another network is called *routing*; it is one of the primary functions of the IP protocol.

Datagrams and Packets

TIPS

IP deals with data in chunks called *datagrams*.

DO YOU KNOW

The terms packet and datagram are often used interchangeably, although a packet is a data link-layer object and a datagram is network layer object.

In many cases, particularly when using IP on Ethernet, a datagram and packet refer to the same chunk of data. There's no guarantee that the physical link layer can handle a packet of the network layer's size. As previously mentioned, the largest packet that can be handled by the physical link layer is called the Maximum Transmission Unit, or MTU, of the network media.

If the medium's MTU is smaller than the network's packet size, then the network layer has to break large datagrams down into packet-sized chunks that the data link and physical layers can digest. This process is called *fragmentation*.

The host receiving a fragmented datagram reassembles the pieces in the correct order. For example, an X.25 network may have an MTU as small as 128 bytes, so a 1518-byte IP datagram would have to be fragmented into many smaller network packets to be sent over the X.25 link. For the scope of this book, we'll use packet to describe both the IP and the data link-layer objects, since NFS is most commonly run on Ethernet rather than over wide-area networks with smaller MTUs. However, the distinction will be made when necessary, such as when discussing NFS traffic over a wide area point-to-point link.

IP Host Addresses

The internet protocol identifies hosts with a number called an *IP address* or a *host address*. To avoid confusion with MAC addresses (which are machine or *station* addresses), the term IP address will be used to designate this kind of address. IP addresses come in two flavors: 32-bit IP Version 4 (IPv4) or 128 bit IPv6 address.

```
    IPv4 addresses are written as four dot-separated decimal
numbers between 0-255 (a dotted quad):
    192.9.200.1
```

IP addresses must be unique among all connected machines. Connected machines in this case are any hosts that you can get to over a network or connected set of networks, including your local area network, remote offices joined by the company's wide-area network, or even the entire Internet community. For a standalone system or a small office that is not connected (via an IP network) to the outside world, you can use the standard, private network addresses assigned such purposes. If your network is connected to the Internet, you have to get a range of IP addresses assigned to your machines through a central network administration authority, via your Internet Service Provider. If you are planning on joining the Internet in the future, you will need to obtain an address from your network service provider. This may be either an actual provider of Internet service, or your own organization, if it has addresses to hand out. We won't go into this further in this book.

The IP address uniqueness requirement differs from that for MAC addresses. IP addresses are unique only on connected networks, but machine MAC addresses are unique in the world, independent of any connectivity. Part of the reason for the difference in the uniqueness requirement is that IPv4 addresses are 32 bits, while MAC addresses are 48 bits, so mapping every possible MAC address into an IPv4 address requires some overlap. There are a variety of reasons why the IPv4 address is only 32 bits, while the MAC address is 48 bits, most of which are historical.

Since the network and data link layers use different addressing schemes, some system is needed to convert or map the IP addresses to MAC addresses. Transport-layer services and user processes use IP addresses to identify hosts, but packets that go out on the network need MAC addresses. The Address Resolution Protocol (ARP) is used to convert the 32-bit IPv4 address of a host into its 48-bit MAC address. When a host wants to map an IP address to a MAC address, it broadcasts an ARP request on the network, asking for the host using the IP address to respond. The host that sees its own IP address in the request returns its MAC address to the sender. With a MAC address, the sending host can transmit a packet on the Ethernet and know that the receiving host will recognize it.

A host can have more than one IP address. Usually this is because the host is connected to multiple physical network segments (requiring one network interface, such as an Ethernet controller, per segment), or because the host has multiple interfaces to the same physical network segment.

IPv4 Address Classes

Each IPv4 address has a *network number* and a *host number*. The host number identifies a particular machine on an organization's network. IP addresses are divided into *classes* that determine which parts of the address make up the network and host numbers, as demonstrated in Table 1-2.

TABLE 2. IPv4 Address Classes,

Address Class and First Octet Value	Network Number Octets	Host Number Octets	Address Form	Number of Networks	Number of Hosts per Network	Maximum Number of Hosts per Class
Class A: 1-126	1	3	N.H.H.H	126	$256^3 - 2$	2,113,928,964
Class B: 128-191	2	2	N.N.H.H	16,384	$256^2 - 2$	1,073,709,056
Class C: 192-223	3	1	N.N.N.H	2,097,152	254	532,676,608
Class D: 224-239	N/A	N/A	M.M.M.M	N/A	N/A	N/A
Class E: 240-255	N/A	N/A	R.R.R.R	N/A	N/A	N/A

Each *N* represents part of the network number and each *H* is part of the address's host number. The 8-bit octet has 256 possible values, but 0 and 255 in the last host octet are reserved for forming broadcast addresses.

Network numbers with first octet values of 240-254 are reserved for future use. The network numbers 0, 127, 255, 10, 172.16-172.31, and 192.168.0-192.168.255 are also reserved :

- 0 is used as a place holder in forming a network number, and in some cases, for IP broadcast addresses.
- 127 is for a host's loopback interface.
- 255 is used for IPv4 broadcast addresses.
- 10, 172.16-172.31, and 192.168.0-192.168.255 are used for private networks that will never be connected to the global Internet.

When the Internet was founded, it was almost impossible to get a class A network number, and few organizations (aside from entire networks or countries) had enough hosts to justify a class A address. Most companies and universities requested class B or class C addresses. A medium-sized company, with several hundred machines, could request several class C network numbers, putting up to 254 hosts on each network. Now that the Internet is much bigger, the rules for class A, B, and C network number assignment have changed.

> **DO YOU KNOW**
>
> **There are only 126 class A network numbers, but well over two million class C network numbers.**

Class D addresses look similar to the other classes in that each address consists of 4 octets with a value no higher than 255 per octet. Unlike classes A, B, and C, a class D address does not have a network number and host number. Class D addresses are multicast addresses, which are used to send messages to more than one recipient host, whereas IP addresses in classes A, B, and C are unicast addresses destined for one recipient. Multicast on the Internet offers plenty of potential for efficient broadcast of information, such as bulk file transfers, audio and video, and stock pricing information, but has achieved limited deployment. There is an ongoing experiment known as the "MBONE" (Multicast backBONE) on the Internet to exploit this technology.

> **DO YOU KNOW**
>
> **Class E addresses are reserved for future assignment.**

Classless IP Addressing

In the early 1990s, due to the advent of the World Wide Web, the Internet's growth exploded. In theory, if you sum the maximum number of hosts per classes A, B, and C , the Internet can have a potential for over 3.7 billion hosts. In reality, the Internet was running out of address capacity for two reasons.

The first had to do with the inefficiencies built into the class partitioning. About 3.2 billion of the theoretical number of hosts were class A and class B, leaving about 500 million class C addresses. Most organizations did not need class A or class B addresses, and of those that did, a significant fraction of their assigned address space was not needed. Most users could get by with a class C network number, but the typical small business or home user did not need 254 hosts. Thus, the number of class C addresses was bounded by the maximum number of class C networks, about two million, which is far less than the number of users on the Internet.

The problem of only two million class C networks was mitigated by the introduction of dynamically assigned IP addresses, and by the introduction of policies that tended to assign

IP network numbers only to Internet Service Providers (ISPs), or to organizations that effectively acted as their own ISP, which would then use the free market to efficiently reallocate the IP addresses dynamically or statically to their customers. Thus most Intenet users get assigned a single IP address, and the ISP is assigned the corresponding network number.

The second reason was routing scalability. When the Internet was orders of magnitude smaller then it is today, most address assignments were for class A or B and so routing between networks was straightforward. The routers simply looked at the network number, and sent it to a router responsible for that route. With the explosion of the Internet, and with most of that growth in class C network numbers, each network's router might have to maintain tables of hundreds of thousands of routes. As the Internet grew rapidly, keeping these tables up to date was difficult.

This situation was not sustainable, and so the concept of "classless addressing" was introduced. With the exception of grandfathered address assignments, each IP address, regardless of whether it's class A, B, or C, would not have an implicit network number part and host number part. Instead the network part would be designated explicitly via a suffix of the form: "/XX", where XX is the number of bits of the IP address that refer to the network. Those organizations that needed more than the 254 hosts that a class C address would provide, would instead be assigned consecutive class C addresses. For example, an ISP that was assigned 192.1.2 and 192.1.3 could have a classless network number of 192.1.3.0/23. Any router on a network other than 192.1.2 or 192.1.3 that wanted to send to either network number would instead route to a single router associated with the classless network number 192.1.3.0/23 (i.e., any IP address that had its first 23 bits equal to 1100 0000 0000 0001 0000 001).

With this new scheme, larger organizations get more consecutive class C network numbers. Within their local networks ("Intranets"), they can either use traditional class-based routing or classless routing that further subdivides the local network address space that can be used. The largest organizations may find that class-based routing doesn't scale, and so classless routing is the best approach.

Virtual Interfaces

It is possible for a physical network segment to support more than one IP network number. For example, a segment might have 128.0.0.0/16 and 192.4.5.6/24. Some hosts on that segment might want to directly address hosts with either network number. Some operating systems, such as Solaris, will let you define multiple virtual or logical interfaces for a physical network interface. On most Unix systems, the *ifconfig* command is used to set up interfaces.

1.3.6. IP Version 6

Until now we have been discussing IPv4 addresses that are four octets long. However, it was recognized long ago, even before the introduction of the World Wide Web, that the IPv4 address space was under pressure. IP Version 6 (IPv6) has been defined to solve the address space limitations by increasing the address length to 128 bit addresses. At the time of this writing, while most installed systems either do not support it or do not use it, most marketed systems support IPv6. Since it seems inevitable that you'll encounter some IPv6 networks in the next few years, we will explain some of the basics of IPv6. Note that IPv6 is sometimes referred to as IPng: IP Next Generation.

Instead of dotted quads, IPv6 addresses are usually expressed as:

```
x:x:x:x:x:x:x:x
```

where each *x* is a 16 bit hexadecimal value. In environments where a network is transitioning from IP Version 4 to Version 6, you might want to use a form like:

```
x:x:x:x:x:x:d.d.d.d
```

where d.d.d.d represents an IP Version 4 dotted quad.

When there are one or more consecutive sequences of *x's* such that each *x* is all zeroes, the sequence can be replaced with "::", but there can be only one such "::" abbreviation in an IPv6 address. Thus:

```
1234:0000:5678:9ABC:DEF0:1234:5678:9ABC
3:0:0:0:0:0:3333:4444
can be abbreviated as:
1234::5678:9ABC:DEF0:1234:5678:9ABC
3::3333:4444
```

As you might expect, IPv6 dispenses with address classes for unicast addresses. You specify classless network numbers (address prefixes), using the same classless addressing notation that IP Version 4 uses.

IP Version 6 Address Pools

While the designation of the network number in IPv6 is classless, the 128-bit address is still carved up into various pools. Portions of the address space are allocated for:

- Reserved or unassigned for future purposes
- Open Systems Interconnection (OSI) network protocols
- Novell IPX protocols
- Unicast addresses, including:
  - (i) global unicast addresses that can be used to send packets to hosts outside the local site
  - (ii) site local unicast addresses than can be used to send packets only to hosts within a site
  - (iii) link local unicast addresses that can used to send packets only to hosts within a physical network segment
- Multicast addresses, which start with FF
- Addresses of nodes that support just IP Version 4. These are denoted as:

```
::FFFF:d.d.d.d
```

- Addresses of nodes that support IPv6, but want to use existing IP Version 4 infrastructure to encapsulate IPv6 packets within IPv4 packets for transport between networks. The last 32 bits of these addresses correspond to IPv4 addresses. These addresses are denoted as:

```
::d.d.d.d
```

While this scheme does not let you benefit from IPv6's extended addressing, it does let you take advantage of IPv6's other features (such as a richer set of protocol options) while transitioning from IPv4.

IP Version 6 Loopback Address

Instead of dedicating about 16 million addresses for loopback interfaces as IPv4 does, IPv6 uses just one address for that purpose:

```
::1
```

IP Version 6 Unspecified Address

IPv6 introduces the concept of an "unspecified" address, which is all zeroes:

```
::0
```

This address can be used by hosts that don't know their own address, but need to generate queries to determine their address assignment. Such hosts would use "::0" as the source address in an IPv6 packet.

Transport Layer

The transport layer has two major jobs: it must subdivide user-sized data buffers into network layer-sized datagrams, and it must enforce any desired transmission control such as reliable delivery. Two transport protocols that sit on top of IP are the Transmission Control Protocol (TCP) and the User Datagram Protocol (UDP), which offer different delivery guarantees.

TCP and UDP

TCP is best known as the first half of TCP/IP; as discussed in this and the preceding sections, the acronyms refer to two distinct services. TCP provides reliable, sequenced delivery of packets. It is ideally suited for connection-oriented communication, such as a remote login or a file transfer. Missing packets during a login session is both frustrating and dangerous — what happens if *rm *.o* gets truncated to *rm ** ? TCP-based services are generally geared toward long-lived network connections, and TCP is used in any case when ordered datagram delivery is a requirement. There is overhead in TCP for keeping track of packet delivery order and the parts of the data stream that must be resent. This is *state* information. It's not part of the data stream, but rather describes the state of the connection and the data transfer. Maintaining this information for each connection makes TCP an inherently *stateful* protocol. Because there is state, TCP can adapt its data flow rate when the network is congested.

UDP is a no-frills transport protocol: it sends large datagrams to a remote host, but it makes no assurances about their delivery or the order in which they are delivered.

UDP is best for connectionless communication on local area networks in which no context is needed to send packets to a remote host and there is no concern about congestion. Broadcast-oriented services use UDP, as do those in which repeated, out of sequence, or missed requests have no harmful side effects.

Reliable and unreliable delivery is the primary distinction between TCP and UDP. TCP will always try to replace a packet that gets lost on the network, but UDP does not. UDP packets can arrive in any order. If there is a network bottleneck that drops packets, UDP packets may not arrive at all. It's up to the application built on UDP to determine that a packet was lost, and to resend it if necessary. The state maintained by TCP has a fixed cost associated with it, making UDP a faster protocol on low-latency, high-bandwidth links. The price paid for speed (in UDP) is unreliability and added complexity to the higher level applications that must handle lost packets.

Port Numbers

A host may have many TCP and UDP connections at any time. Connections to a host are distinguished by a *port number*, which serves as a sort of mailbox number for incoming datagrams. There may be many processes using TCP and UDP on a single machine, and the port numbers distinguish these processes for incoming packets. When a user program opens a TCP or UDP socket, it gets connected to a port on the local host. The application may specify the port, usually when trying to reach some service with a well-defined port number, or it may allow the operating system to fill in the port number with the next available free port number.

When a packet is received and passed to the TCP or UDP handler, it gets directed to the interested user process on the basis of the destination port number in the packet. The quadruple of:

```
source IP address, source port, destination IP address,
destination port
```

uniquely identifies every interhost connection in the network. While many processes may be talking to the process that handles remote login requests (therefore their packets have the same destination IP addresses and port numbers), they will have unique pairs of source IP addresses and port numbers. The destination port number determines which of the many processes using TCP or UDP gets the data.

On most Unix systems port numbers below 1024 are reserved for the processes executing with superuser privileges, while ports 1024 and above may be used by any user. This enforces some measure of security by preventing random user applications from accessing ports used by servers. However, given that most nodes on the network don't run Unix, this measure of security is very questionable.

The Session and Presentation Layers

The session and presentation layers define the creation and lifetime of network connections and the format of data sent over these connections. Sessions may be built on top of any supported transport protocol — login sessions use TCP, while services that broadcast information about the local host use UDP. The session protocol used by NFS and NIS is the Remote Procedure Call (RPC).

The Client-Server Model

RPC provides a mechanism for one host to make a procedure call that appears to be part of the local process but is really executed on another machine on the network. Typically, the host on which the procedure call is executed has resources that are not available on the calling host. This distribution of computing services imposes a client/server relationship on the two hosts: the host owning the resource is a server for that resource, and the calling host becomes a client of the server when it needs access to the resource. The resource might be a centralized configuration file (NIS) or a shared filesystem (NFS).

Instead of executing the procedure on the local host, the RPC system bundles up the arguments passed to the procedure into a network datagram. The exact bundling method is determined by the presentation layer, described in the next section. The RPC client creates a session by locating the appropriate server and sending the datagram to a process on the server that can execute the RPC; see Fig. 1. On the server, the arguments are unpacked, the server executes the result, packages the result (if any), and sends it back to the client. Back on the client side, the reply is converted into a return value for the procedure call, and the user application is re-entered as if a local procedure call had completed. This is the end of the "session," as defined in the ISO model.

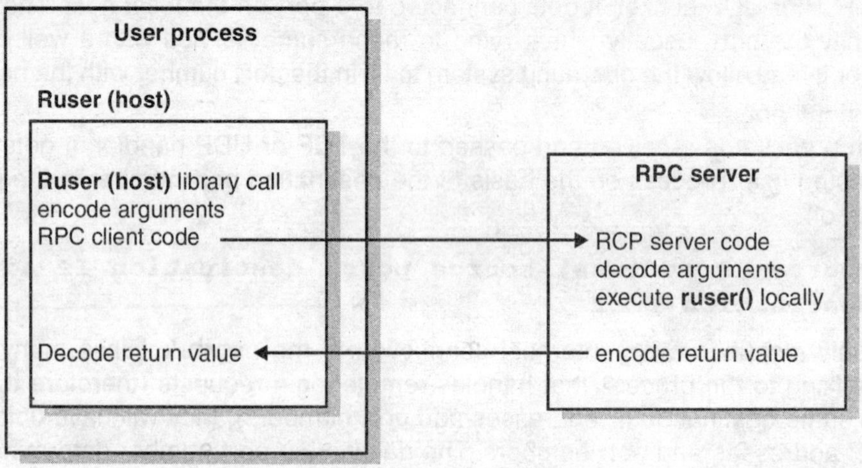

Fig. 1. Remote procedure call execution

RPC services may be built on either TCP or UDP transports, although most are UDP-oriented because they are centered around short-lived requests. Using UDP also forces the RPC call to contain enough context information for its execution independent of any other RPC requests, since UDP packets may arrive in any order, if at all.

When an RPC call is made, the client may specify a timeout period in which the call must complete. If the server is overloaded or has crashed, or if the request is lost in transit to the server, the remote call may not be executed before the timeout period expires. The action taken upon an RPC timeout varies by application; some resend the RPC call, while others may look for another server.

External data representation

At first look, the data presentation layer seems like overkill. Data is data, and if the client and server processes were written to the same specification, they should agree on the format of the data — so why bother with a presentation protocol? While a presentation layer may not be needed in a purely homogeneous network, it is required in a heterogeneous network to unify differences in data representation. These differences are outlined in the following list:

Data byte ordering

Does the most significant byte of an integer go in the odd- or even-numbered byte?

Compiler Behavior

Do odd-sized quantities get padded out to even-byte boundaries? How are unions handled?

Floating Point Numbers

What standard is used for encoding floating point numbers?

Arrays and Strings

How do you transmit variable-sized objects, such as arrays and strings?

Again, a presentation protocol would not be necessary if datagrams consisted only of byte-oriented data. However, applications that use RPC expect a system call-like interface, including support for structures and data types more complex than byte streams. The presentation layer provides services for encoding and decoding argument buffers that may then be passed down to RPC for transmission to the client or server.

The External Data Representation (XDR) protocol was developed by Sun Microsystems and is used by NIS and NFS at the presentation layer. XDR is built on the notion of an immutable network byte ordering, called the *canonical form*. It isn't really important what the canonical form is — your system may or may not use the same byte ordering and structure packing conventions. The canonical form simply allows network hosts to exchange structured data (as opposed to streams of bytes) independently of any peculiarities of a particular machine. All data structures are converted into the network byte ordering and padded appropriately.

The rule of XDR is "sender makes local canonical; receiver makes canonical local." Any data that goes over the network is in canonical form.A host sending data on the network converts it to canonical form, and the host that receives the data converts it back into its local representation. A different way to implement the presentation layer might be "receiver makes local." In this case, the sender does nothing to the local data, and the receiver must deduce the packing and encoding technique and convert it into the local equivalent. While this scheme may send less data over the network — since it is not subject to additional padding — it places the burden of incorporating a new hardware architecture on the receiving side, rather than on the new machine. This doesn't seem like a major distinction, but consider having to change all existing, fielded software to handle the new machine's structure-packing conventions. It's usually worth the overhead of converting to and from canonical form to ensure that all new machines will be able to "plug in" to the network without any software changes.

The canonical form matches the byte ordering of the Motorola and SPARC family of microprocessors, so these processors do not have to perform any byte swapping to translate to or from canonical form. This byte ordering is called Big Endian. Big Endian ordering is used for many Internet protocols.

The XDR and RPC layers complete the foundation necessary for a client/server distributed computing relationship. NFS and NIS are client/server applications, which means they sit at the top layer of the protocol stack and use the XDR and RPC services. To complete this introduction to network services, we'll take a look at the two mechanisms used to start and maintain servers for various network services.

Internet and RPC server configuration

The XDR and RPC services are useful for applications that need to exchange data structures over the network. Each new RPC request contains all required information in its XDR-encoded arguments, just as a local procedure call gets its inputs from passed-in arguments. RPC services are usually *connectionless* services because RPC requests do not require the creation of a long-lived network connection between the client and server. The client communicates with the server to send its request and receive a reply, but there is no connection or environment for the communication.

There are many other network services, such as

- telnet
- ftp,

that are commonly referred to as the Internet or ARPA services. They are part of the original suite of utilities designed for use on the Internet. Internet services are generally based on the TCP protocol and are *connection-oriented* — the service client establishes a connection to a server, and data is then exchanged in the form of a well-ordered byte stream. There is no need for RPC or XDR services, since the data is byte-oriented, and the service defines its own protocols for handling the data stream. The *telnet* service, for example, has its own protocol for querying the server about end-of-line, terminal type, and flow control conventions.

> **DO YOU KNOW**
>
> **RPC services are not required to be connectionless.**

RPC can be run over TCP, in a connection-oriented fashion. The TCP transport protocol may be used with RPC services whenever a large amount of data needs to be transferred. NIS, for example, uses UDP (in connectionless mode) for most of its operations, but switches to TCP whenever it needs to transfer an entire database from one machine to another. NFS supports either TCP or UDP for all its operations.

Most Internet services are managed by a super-daemon called *inetd* that accepts requests for connections to servers and starts instances of those servers on an as-needed basis. Rather than having many server processes, or *daemons*, running on each host, *inetd* starts them as requests arrive. Clients contact the *inetd* daemon on well-known port numbers for each service. These port numbers are published in the */etc/services* file.

inetd *sets up a one-to-one relationship between service clients and server-side daemons. Every* rlogin *shell, for example, has a client side* rlogin *process (that calls* inetd *upon invocation) and a server-side* in.rlogind *daemon that was started by* inetd. *In this regard,* inetd *and the services it supports are* multi-threaded*: they can service multiple clients at the same time, creating a new separate connection (and state information) for each client. A new server instance, or thread, is initiated by each request for that service, but a single daemon handles all incoming requests at once.*

Only traffic specific to a single session moves over the connection between a client and its server. When the client is done with the service, it asks the server to terminate its connection, and the server daemon cleans up and exits. If the server prematurely ends the connection due to a crash, for example, the client drops its end of the connection as well.

Some RPC services can't afford the overhead of using *inetd*. The standard *inetd*based services, like *telnet*, tend to be used for a long time, so the cost of talking to *inetd* and having it start a new server process is spread out over the lifetime of the connection. Many RPC calls are short in duration, lasting at most the time required to perform a disk operation.

RPC servers are generally started during the boot process and run as long as the machine is up. While the time required to start a new server process may be small compared to the time a remote login or *rsh* session exists, this overhead is simply too large for efficient RPC operation. As a result, RPC servers typically have one server process for the RPC service, and it executes remote requests for all clients in the same process. Some RPC servers are single-threaded: they execute requests one at a time. To achieve better performance, some RPC servers are multi-threaded: they have multiple threads of execution within the same process, sharing the same address space. There may be many clients of the RPC server, but their requests intermingle in the RPC server queue and are processed in the order in which server threads are dispatched to deal with the requests.

Instead of using pre-assigned ports and a super-server, RPC servers are designated by service number. The file */etc/rpc* contains a list of RPC servers and their program numbers. Each program may contain many procedures. The NFS program, for example, contains more than a dozen procedures, one for each filesystem operation such as "read block," "write block," "create file," "make symbolic link," and so on. RPC services still must use TCP/UDP port numbers to fit the underlying protocols, so the mapping of RPC program numbers to port numbers is handled by the portmapper daemon (*portmap* on some systems, *rpcbind* on others).

When an RPC server initializes, it usually registers its service with the portmapper. The RPC server tells the portmapper which ports it will listen on for incoming requests, rather than having the portmapper listen for it, in *inetd* fashion. An RPC client contacts the portmapper daemon on the server to determine the port number used by the RPC server, or it may ask the portmapper to call the server indirectly on its behalf. In either case, the first RPC call from a client to a server must be made with the portmapper running. If the portmapper dies, clients will be unable to locate RPC daemons services on the server. A server without a running portmapper effectively stops serving NIS, NFS, and other RPC-based applications.

Socket RPC and Transport Independent RPC

RPC was originally designed to work over sockets, a programing interface for network communication introduced in the 1980s by the University of California in its 4.1c BSD version of Unix. Solaris 2.0 introduced Transport Independent RPC (TI-RPC). The motivation for TI-RPC was that it appeared that OSI networking would eventually supplant TCP/IP-based networking, and so a transport independent interface would make it easier to transition RPC applications was needed. While OSI networking did not take over, TI-RPC is still used in Solaris. TI-RPC introduces an additional configuration file, */etc/netconfig*, which defines each transport that RPC services can listen for requests over. In addition to TCP and UDP, the */etc/netconfig* file lists connectionless and connection-oriented loopback transports for

RPC services that don't need to provide service outside the host. In Solaris 8, the */etc/netconfig* file will also let you specify services over TCP and UDP on IPv6 network interfaces.

Purpose of Directory Services

The purpose of a directory service is to map names of one form to names of another form. Often the names of the first form are alphanumeric strings, and the second form are numbers. Or the names of the first form are simple, whereas the names of second form are complex. In the days before computing, we used directories, but they were published on paper. The most obvious one, and perhaps the inspiration for network directory services, is the telephone book. A typical telephone book for a city in the United States consists of three directories:

- The white pages of residence listings
- The white pages of business listings
- The yellow pages of business listings

The residence white page listings contains a list of names, last name first, and for each entry, the telephone number. In many cases, each entry contains the street address of the residence. Thus you can think of the residence white pages in a telephone book, as a way to *direct* you from a person's name to his telephone number and address. Hence, a telephone book's formal name is a telephone *directory*.

The white pages of residence listings can be thought of as a "structured set of data." If this data were stored on a computer, a lexicographer would call it a *database* ; a "structured set of data" is the definition of *database* in the Concise Oxford Dictionary. Sometimes, when you are in a hurry or don't have a telephone number, you dial a special telephone number (411 or 555-1212 in the United States) to ask an operator ("directory assistance") for the telephone number of the person you want to call. This directory assistance can be thought of as directory service. You, the caller, are a customer or client of the directory service, and the particular operator, is a server of the directory service. In the world of computer networking, the human server is replaced with a directory server of databases. Since there is a server, there has to be a client. The client-side of a directory service is typically a programming library which allows other applications to look up entries in the database.

The Hosts Database

The metaphor of a telephone directory was useful in explaining the concepts of directory service, client, and server. It turns out that the concept of names of people and their telephone numbers is also a metaphor for a similar database in computer networking. hosts have unique numbers or addresses, just as every telephone number in the world is unique. Just as we associate names of people with their telephone numbers, in computer networking we often want to give individual hosts a name in addition to a host address. The reasons are that it is easier to remember a name than a number, and just as people move geographically, requiring new telephone number assignments, hosts can move physically (requiring a new address assignment for the host) or conversely, the function the host was serving can move from one physical host to another (requiring a new name assignment for the host).

The hostname and address entries are stored in a hosts database that the directory server can use to respond to requests from clients. As was noted earlier, the client-side of the directory service is typically a programming library. This is the case for the hosts database.

There is a subroutine, known as *gethostbyname( )* that takes a string name of a host and returns the address of the host. See your system's manual page for *gethostbyname* for the precise calling conventions. Solaris comes with a utility called *getent* for looking up database entries via the command line. For example:

```
% getent hosts frostback
128.0.0.1          frostback
```

getent *can be thought of as one of the most primitive directory service clients, but nearly every application that deals with the network will be a client that needs to access the hosts database via the directory service. A more advanced client of the hosts database is a web browser such as Netscape Navigator or Internet Explorer. Browsers will link to* gethostbyname *or a similar interface to find the host addresses corresponding to Universal Resource Locators (URLs, those things that start with* http:// *).*

Going back to the telephone concept, sometimes we would like to know the name of the caller corresponding to a telephone number. In the United States, when you call a toll free number, the merchant receiving the call has the capability to display your phone number and can map it to your name (considering that the merchant is paying for your long distance call, some might reason that this is fair). In the computer networking world, it is sometimes useful to know the hostname of the client accessing the server. For example, suppose the server side of a web browser is a web server. Web servers often keep logs of the "hits" made to the server, for the purposes of understanding how popular a web site is, what is popular, and what hosts find it popular. The web server will always be able to find the host address of the client that made the hit. To figure out the name of the host, there is a programming interface called *gethostbyaddr( )*, which takes a host address, and returns the name of the host. The information can be obtained from the hosts database, via the directory service. In other words, both servers and client of different services, in this case, web services, can be clients of directory services.

Directory Name Service (DNS)

The roots of DNS are in the early (pre-Web) days of the Internet. DNS was developed to provide hostname and address resolution. Before DNS existed, the authorities for the Internet maintained a global flat text file of the mappings from hostname to IP address in a file called *hosts.txt*, which was then made available for all the nodes on the Internet to download via a program called *gettable*. This is analogous to the telephone company giving you an updated telephone book periodically. Systems like Unix would convert the file into */etc/hosts*.

This *hosts.txt* system worked fine for the Internet when it had only thousands of hosts. But when it reached tens of thousands of hosts, it wasn't practical, especially over the slow links available in the late 1980s. What was needed was a way to decentralize the process of looking up hostnames and addresses. The Internet was separated into domains, and each domain was left to identify its own authoritative server for hostnames and addresses within its domain. The only thing that needed to be maintained in a global database was the list of domain names, and the servers for that domain. Returning to the telephone directory analogy, when you live in one area code of the United States, and want to get directory information for another area code, you can prefix the area code to the number 555-1212 to get the appropriate directory service operator.

By assigning authority for a domain's directory information to each domain, DNS can be described as being hierarchical. Similarly, the United States telephone system assigns authority for a given area code's directory information to one pool of directory service operators that answer the 555-1212 number. DNS also lets domains within subdomains further delegate authority, and subdomains in turn. For example, in DNS there is a top-level domain called ".com" that assigns authority for administering *sun.com* and *oreilly.com* to DNS servers that the owners of sun.com and oreilly.com each designate. Within *sun.com*, there are several subdomains, such as *eng.sun.com*, and *east.sun.com*. Within *eng.sun.com*, there might be a *compiler.eng.sun.com*, *sunos.eng.sun.com*, *cde.eng.sun.com*. Thus DNS is a multilevel hierarchy, whereas the United States telephone directory service has but two levels of hierarchy.

DNS has stood the test of time. In 1993, a memorandum (RFC 1401) was written by the chair of the Internet Architecture Board that noted that the transition from hosts.txt to DNS was largely complete. This is fortuitous, as the World Wide Web was about to explode from tens of thousands of hosts to millions. DNS proved capable of handling that explosion.

Network Information Service (NIS)

NIS was developed by Sun Microsystems in the mid-1980s to solve a problem that until then had no solution in the Unix world. Let's return to the telephone directory service concept. One nice thing about calling your telephone company's directory service is that the operator (the server) is more apt to have up-to-date information than you would. Your telephone book is replaced once a year, whereas the server's information is updated more frequently, perhaps instantly with each new telephone number assignment and de-assignment. When networking was added to Unix systems, system administrators very quickly ran into difficulties keeping files like */etc/hosts* (holds hostname to host address mappings) and */etc/passwd* (holds username, user identifier, password). If a system administrator had 100 systems, then adding a host to a network or a user to the organization meant the tedium of updating the */etc/hosts* or */etc/passwd* files on all 100 systems. NIS, originally called the Yellow Pages or YP, was invented to simplify management of these files by changing the underlying programming interfaces, such as *gethostbyname( )* and *getpwnam( ),* to use NIS client libraries.

While DNS was being developed around the time NIS was, DNS was mostly concerned with the directory of hostnames and addresses, whereas NIS went beyond that. In addition, DNS was designed so that a host in one domain could access information from other domains, whereas NIS shared the limitations of the early Internet's *hosts.txt* file: flat and not very dynamic.

NIS+

In 1992, Sun Microsystems released NIS+ with Solaris 2.0. Despite its name, NIS+ was more different than it was similar to NIS. NIS+ was developed to address several deficiencies in NIS:

Hierarchical Operation

While NIS was designed to be split into unique domains, there was no simple way for a client in one domain to get directory information from another domain. NIS+ addressed this by supporting a multilevel hierarchy in a manner similar to DNS.

Security

There are really two issues here. First is that some kinds of directory information need to be kept more secure than others, such as a directory containing credit card numbers. The directory server needs to know who is accessing the data, and properly authenticate the client. Second, the client needs to be certain that the server is the true authority for the service. An attacker in the middle between the client and real server could masquerade as the server and return bogus information. NIS+ deals with both these issues by supporting mutual authentication: the client and server authenticate each other, via a secure form of RPC known as RPC/dh,

Updates

Updating a NIS database and propagating the changes is a cumbersome process. Only the system administrator can make updates (with few exceptions), and the changes must be pushed to each replica server by pushing the entire database, even if only one record changes. NIS+ supports the ability to allow users to update directory entries they have access rights to. For example, a user changes the name that appears in the password database, which might be necessary upon a status change like a new job title, or a new surname as a result of a marriage or divorce. NIS+ servers have the capability to accept incremental updates, which allows the updates to be more efficiently distributed.

X.500

Around the same time DNS and NIS were being designed and deployed, the International Standards Organization (ISO) started meeting to define an ISO standard directory, called X.500. X.500 shares DNS's and NIS+'s attributes for hierarchical operation, and NIS+'s attributes for security and simple update. X.500 differs from DNS, NIS, and NIS+ in the following ways:

·X.500 is very explicit about what each level of the hierarchy of a domain name looks like. If you see a domain name like:

chicago.manufacturing.widget.com

it could easily be a DNS, NIS, or NIS+ name, and it could very well be for a host or a domain. Moreover, while it might seem like *chicago.manufacturing.widget.com* refers to a subdomain of hosts located in the city of Chicago, assigned to Widget, Inc.'s manufacturing division, it could just as easily refer to a hand held computer on space station Alpha. This ambiguity is a concern to some folks. Hence, X.500 explicitly identifies what each level of hierarchy means. For example, the X.500 distinguished name corresponding to DNS style *chicago.manufacturing.widget.com* name would be:

{ Country = US, Orglanization = Widget, Inc., Organizational Unit =
Manufacturing, Location = Chicago }

·X.500 supports the notion of *schema*. A schema is a set of rules for what can be stored in a database. Defining a directory schema is useful when performing search operations on a directory. Say a database includes the hire dates of employees, and you want to search for all employees hired between a particular range of dates. Because the X.500 directory is "aware" that the field being searched is a date, it is possible to let the directory server do all the work of finding the matches. With DNS, NIS, and NIS+, you would be compelled to read every directory entry from the server, and perform the operation on the client, because the server treats the information opaquely. The X.500 way saves network bandwidth.

For many common databases, X.500 is overkill, but there are situations where having an X.500 schema is handy. Say you want to find all the hosts that are in the 192.1.1 network. If you defined X.500's equivalent to the hosts database with a schema that had substring matching rules, this would be easy and efficient.

Lightweight Directory Access Protocol (LDAP)

X.500 has a protocol called the Directory Access Protocol (DAP) to allow clients to access X.500 servers. DAP was designed to operate over ISO's Open Systems Interconnect (OSI) transport and network protocols. Once upon a time, people believed that TCP/IP would wither away and be replaced by OSI. As it turned out, too many people had deployed TCP/IP-based networks, and they saw no compelling reason to switch to OSI. Despite OSI mandates by most governments in the developed world, the Internet transport and network protocols persisted, and it was obvious by 1994, if not earlier, that the OSI transport and network was dead. However, as discussed earlier, X.500 has some extremely attractive properties for a directory, but it comes with the baggage of OSI transport and complex ASN.1 encoding. The Lightweight Directory Access Protocol (LDAP) was invented to allow clients using TCP/IP and simpler encoding schemes, to take advantage of the richness of X.500 directory service.

Another difference between LDAP and DAP is that LDAP is under the control of the Internet Engineering Task Force (IETF), the same organization that produced the standards behind the Internet. Whether intended or not, the effect is to get IETF to buy into X.500, whereas previously IETF had no control over OSI transport and network, and so it was much harder (and eventually impossible) to get IETF to accept OSI transport and network.

LDAP specifies lots of different security flavors, including ones based on public key certificates and Kerberos V5.

At the time this book was written, LDAP was only starting to be integrated with operating systems. Windows 2000 is the first such offering from Microsoft. Solaris 8 includes a fully integrated LDAP client, but no server.

NT Domain

NT Domain is the directory service used in Windows NT. It was introduced by Microsoft in 1987 and was called Lan Manager at the time. NT Domain is intended to administer users, groups, printers, and hosts in a Windows environment. NT Domain now supports multilevel hierarchies, but requires a bilateral trust relationship between each domain. So if there are N domains in an organization, N * (N - 1) relationships need to be set up. NT Domain supports slightly better security than NIS. Perhaps the biggest issue with NT Domain is that it is an undocumented proprietary protocol, making it difficult for Windows and non-Windows systems to share NT Domain directory information.

Microsoft is moving away from NT Domain in favor of Active Directory, which is a derivation of the LDAP protocol and X.500.

While NT Domain is not supported on Solaris and most other Unix systems, if you have a mixed environment, you'll probably run into it

Name Service Switch

With multiple directory services available, having the ability to access different ones is handy. Solaris has an */etc/nsswitch.conf* file that for each database, which lets you decide what directory you want to get the database contents from. You can even specify multiple directories. For example, *nsswitch.conf* might have this entry:

```
hosts:    files nis dns
```

This entry says that when *gethostbyname*( ) and *gethostbyaddr( )* are called to look up hostnames and addresses, the interfaces will first try to find the information in the local */etc/hosts* file, then check with NIS, then check with DNS. Be aware that some directory services can't be combined in *nsswitch.conf*. For example, you cannot have both NIS and NIS+ listed in *nsswitch.conf,* even for different databases.

Which directory service to use

Clearly, LDAP is the future for directory services on all operating systems, including Solaris. However, at the time this book was written, LDAP was only starting to be integrated with operating systems. Windows 2000 is the first such offering from Microsoft. Solaris 8 includes a fully integrated LDAP client, but no server. Moreover, LDAP is more complex to administer than other directory services.

NIS is perhaps the easiest to administer, but it is also the most limited. It is, however, the universal directory for Unix systems.

DNS is the standard for hostnames and addresses, and you'll find it handy for accessing hosts outside your domain.

NIS+ has gained some acceptance among other non-Solaris Unix operating systems, including HP's HP-UX, IBM, AIX, and Linux. NIS+ is much more secure than NIS.

This rest of this book ignores NIS+ and LDAP, and focuses on NIS and to some degree DNS, since those are what you are most likely to encounter. If you are concerned about security, you'll need to seriously consider deploying NIS+ until LDAP catches up. If security is not a concern, then NIS is fine

Domain Name Servers

The hostname management provided by NIS can be integrated with an Internet Domain Name Service (DNS), or the DNS facilities can be used to replace the NIS host map in its entirety. We'll avoid a full-length discussion of setting up a name server. That process depends on the type of name server supported by your vendor, and it is best described by your vendor's documentation. Instead, this section concentrates on differences between the scope of the two hostname services, and support for DNS with and without NIS. Note that the implementation of Domain name services provided by your vendor may not be called DNS. If the Berkeley Internet Name Domain name service or one of its derivatives is used, the service is often called BIND.

DNS versus NIS

DNS provides a hierarchical hostname management system that spans the entire Internet. Each level in the hierarchy designates authoritative name servers that contain maps of hostnames and IP addresses, similar to the NIS hosts map but on a larger scale. The DNS server for a large name service domain would have host information merged from dozens of NIS domains. First among the advantages of DNS is its ability to decentralize responsibility for the maintenance of hostname-to-IP address mappings and the resulting domain name qualification that is used to differentiate identically named hosts.

Decentralized name management means that each organization running a name service domain — whether it is a subdivision of a corporation or an entire company — can maintain its own host information without having to notify some central authority of changes in its local configuration. Host information is published through the authoritative name server for that domain, and hosts in other name service domains retrieve information from the name server when needed. Every domain knows how to reach the next highest level in the name space hierarchy, and it can generally find most of its peer name servers within the same organization. If a name server does not know how to reach the name server for another domain, it can ask the next higher level domain name server for assistance.

For example, Princeton University is part of the educational, or *.edu,* domain. The domain name for the entire university is *princeton.edu*, and it is further divided by department:

```
cs.princeton.edu
politics.princeton.edu
history.princeton.edu
```

and so on. Each of the name servers for the departmental name service domains knows how to reach most of the others; therefore each department can run its own systems without having to notify a campus-wide network manager of any changes to host information. There is also a name server for the entire *princeton.edu* domain that points to lower-level name servers for incoming queries and locates other domains in *.edu*, *.com*, or *.gov* for outbound requests.

In a world in which every machine name must be unique, all of the good names are taken very quickly. DNS allows each domain to have a distinct name space, so that two domains may have hosts with the same name: the name service domain suffix distinguishes them on a higher level in the hierarchy. This is a job that cannot be performed by NIS, since the concatenation of */etc/hosts* files from several different domains would result in hostname clashes. If the NIS domains are left independent, there is no global naming authority, because NIS lacks a mechanism for cross-domain hostname queries.

DNS Integration with NIS

Hostnames are managed in a hierarchy. Each host manages its own name, so the hosts are the "leaf nodes" in this management tree. Hosts are grouped together into NIS or DNS domains, creating a two-level tree. DNS domains may be further grouped together by company, department, or physical location, adding more levels to the management hierarchy. NIS fits into the DNS management scheme at the lowest level in this hierarchy.

Within a single DNS domain, there may be many physical networks with several system administrators. NIS provides a system for the independent management of these small networks; NIS host map information can be combined to form the DNS host file.

NIS and DNS domain names

If an Internet DNS is used in conjunction with NIS, it is helpful to tie the NIS domain names to the DNS domain name. Deriving NIS domain names from the DNS domain name links the two management schemes: the DNS-derived portion of the NIS domain name indicates where the NIS domain looks for its hostname information. Joining NIS and DNS domain names also makes sense if you have a single DNS domain that spans several physical locations. Each office will have its own networks, and its own NIS domains, so using the DNS domain name in the NIS domain name indicates how these locations fit into the "big picture."

For example, the Polygon Company uses the DNS domain name *polygon.com*. It has four NIS domains in its main office, which uses the *polygon.com* DNS domain name. The NIS domain names use the DNS domain name as a suffix:

```
bos-engin.polygon.com
philly-engin.polygon.com
finance.polygon.com
sales.polygon.com
```

If NIS is set up as the primary directory service, then Solaris versions of *sendmail* assume that an NIS domain name was derived from a DNS domain name, and they will strip the first component to derive the mail domain name. That is, if your NIS domain name is *bos-engin.polygon.com*, then *sendmail* uses *polygon.com* as your mail domain name by default. There may be many NIS domains in this DNS domain; *sendmail* strips off the leading component to form the DNS domain name.

However, if there are multiple NIS domains within the DNS domain — several sales offices in different cities, for example — then the NIS domain names should reflect the subdivision of the DNS domain, as shown in Table 3.

TABLE 3. Subdividing a DNS domain into NIS domains

NIS Domain	DNS Domain
boston.sales.polygon.com	.sales.polygon.com
philly.sales.polygon.com	.sales.polygon.com
rahway.sales.polygon.com	.sales.polygon.com
waltham.engin.polygon.com	.engin.polygon.com
alameda.engin.polygon.com	.engin.polygon.com

Because the NIS domain name contains four dot-separated components, *sendmail* drops the first component and uses the remainder as a DNS domain name. This allows all of the sales offices to be treated as a single administrative unit for mail and hostname management, even though they require distinct NIS domains.

It is important to note that each single administrative unit, whether it is implemented with one NIS domain or multiple NIS domains must share the same map entries. Thus, all the hosts listed in the hosts map of *waltham.engin.polygon.com* must be listed in the hosts map of *alameda.engin.polygon.com*. The converse must be true as well. Getting all hosts to agree on usernames, *uid/gid* values, and host addresses is a prerequisite for adding other distributed services such as the Network File System.

Domain Aliases

Some systems impose a fairly small limit on the length of a domain name. If you've chosen a long NIS domain name, say *nesales.East.Sun.COM*, then implementations of NIS that restrict the length of a domain name will not be able to bind to a server.

You could build a second NIS domain with a shorter name and duplicate the maps from the first domain, but this leaves you with twice the administrative work. An easier solution to this problem is to create a domain name alias for the longer name by making a symbolic link in the NIS server directory */var/yp* on each server host:

```
master# cd /var/yp

master# ln -s nesales.East.Sun.COM nesales
```

NIS servers in the fully qualified domain respond to requests for service for the truncated domain name because they believe they have a set of maps for the specified domain. It is of no consequence that the "directory" is really a link to another domain's directory. This trick can also be used to force two distinct NIS domains to share exactly the same set of maps.

In a simple network, your domain names are likely to be short and easily managed. However, if you integrate DNS with NIS, and choose NIS domain names based on name service domains, you may end up with long, multicomponent names such as *grad.history.princeton.edu*. Using symbolic links to create aliases for long names may be necessary to make all of your NIS clients find NIS servers.

There are four ways to integrate NIS with DNS, each of which is described in more detail in the following subsections.

Run NIS without DNS on Client and Server

This is the default for many systems, including Solaris. In this approach, the name services switch file, *nsswitch.conf*, is set up so that *nis* and *files* are the only directory services listed in the *hosts* entry of the *nsswitch.conf* file. The NIS server is configured (by default) to not use DNS to resolve hostnames not found in the hosts map.

Run NIS on Client, Enable DNS on NIS Server

Use the NIS maps first, then go to DNS for hostnames that aren't managed by NIS. This is done using a special flag in the NIS hosts map.

NIS is forced to query DNS for hostnames not found in the hosts map if the map is built with the "Inter-Domain" key. The NIS-then-DNS algorithm is embedded in the implementation of *ypserv*. This means that individual NIS clients don't need to know about the DNS; only the NIS servers will be calling DNS for non-local hostnames.

In the NIS *Makefile*, add the *-b* flag to the *makedbm* script for the *hosts.byname* and *hosts.byaddr* maps, which will cause the *YP_INTERDOMAIN* key to be added to the *hosts* maps. In Solaris, this is done by changing the following lines in */var/yp/Makefile* from:

```
#B=-b
B=
to:
B=-b
#B=
```

If a hostname is not found in the NIS map, the *YP_INTERDOMAIN* keyinstructs NIS to look up the name with the domain name server. Instead of immediately returning an error indicating that the hostname key was not found, *ypserv* asks the DNS server to look up the hostname. If DNS cannot find the name, then *ypserv* returns an error to the client. However, if the DNS server locates the hostname, it returns the IP address information to *ypserv*, and *ypserv* returns it to the client. Integration of NIS and DNS is completely invisible to the client in terms of calling interfaces: all of the work is done by *ypserv* on the NIS server.

NIS servers locate DNS servers through the *resolver* interface, which relies on information in the */etc/resolv.conf* configuration file. The resolver configuration file should point to at least two DNS servers to provide redundancy in case one DNS server becomes unavailable:

```
nameserver 130.1.52.28
nameserver 130.1.1.15
```

The *nameserver* keyword is used to identify the IP address of a DNS server. The servers are listed by IP address, since hostnames are dependent on the very mechanism being configured by this file. Set up a *resolv.conf* file on every NIS server.

Run DNS on NIS Clients and Servers

In this approach, NIS clients and servers ignore NIS for hostnames and use only DNS.

Given that DNS is a full-service hostname management system, some network managers choose to eliminate the NIS hosts map file and use pure DNS service for hostnames and IP addresses. On some systems, a new version of *gethostent*( ) is required that skips the NIS query and directly calls the DNS *resolver* routines for hostname lookups. Check with your vendor for instructions on how to do this. In Solaris, the name services switch can be used to set the hosts lookups to just DNS. If you disable NIS hostname management and use DNS alone, you'll need to set up a *resolv.conf* file on *every* host in the network, so that they can find DNS servers.

The main argument for using DNS only is that it consolidates hostname management under one distributed service, instead of having it split across two services. The drawbacks to this approach are that each host is then dependent upon both an NIS and a DNS server for normal operation (if NIS is running), and a reliable DNS server or sufficient resolver information is required to make each small network self-supporting. Widespread use of DNS to replace NIS host maps suffers from the same server availability problems that NIS does — the entire network is dependent upon reliable and well-behaved servers.

Run NIS on client, enable DNS on NIS client

In this approach, the name services switch file, *nsswitch.conf,* is set up so that both *nis* and *dns* appear in the *hosts* entry of the *nsswitch.conf* file. The host maps should not have the "Inter-Domain" key enabled since all that will do is result in hostname resolutions via DNS occurring twice: once in the NIS client and once in the NIS server.

Centralized Versus Distributed Management

If your organization has centralized remote control of all soft administration, then these issues will be of less interest to you.

Soft administration includes everything that does not require onsite personnel. An example of something that is not soft administration would be replacing a disk drive.

NIS lends itself to allowing you to give system administration groups for a given department within your organization responsibility for maintaining the department's NIS maps without the need for centralized control. However, the nature of hostnames, host addresses, and domain name management is that some central controls or rules are necessary in order to prevent mistakes in one department from affecting other departments and beyond.

There are at least three basic approaches to consider for managing hosts and domains.

Complete centralization

In this model, if someone wants an IP address, he or she contacts a single central committee to get one; the chances of errors are as low as possible, but the latency in getting requests honored is the longest. Adding new subdomains is also centralized. In this model, as there are specific system management groups managing the non-hosts NIS maps for a given department, it is not practical to manage hosts via NIS; you would use DNS exclusively.

Federation

In this model, the central committee has delegated responsibility for portions of the IP address space to individual groups responsible for a DNS subdomain. In this model, either a DNS or a hybrid NIS/DNS model for managing hosts works. If the individual groups are using DNS to the exclusion of the NIS hosts map, then there is little work for the central committee other than to maintain the mapping of subdomains to subdomain name servers. The central committee, of course, is responsible for adding or deleting subdomains. If the individual groups use NIS for local hostname information, then the central committee would maintain the entire DNS infrastructure by periodically gathering host map information from each group. This could be done automatically.

Complete decentralization

Each system administration group has the autonomy to modify its NIS host maps as well as the authority to modify the common DNS database. Such a system will not scale as the number of subdomains and system administration groups rises. With too many authorized players, it will be hard to track down problems caused by mistakes, not to mention avoiding duplicate efforts.

NFS

The Network File System (NFS) is a distributed filesystem that provides transparent access to remote disks. Just as NIS allows you to centralize administration of user and host information, NFS allows you to centralize administration of disks. Instead of duplicating common diréctories such as */usr/local* on every system, NFS provides a single copy of the directory that is shared by all systems on the network. To a host running NFS, remote filesystems are indistinguishable from local ones. For the user, NFS means that he or she doesn't have to log into other systems to access files. There is no need to use *rcp* or tapes to move files onto the local system. Once NFS has been set up properly, users should be able to do all their work on their local system; remote files (data and executables) will appear to be local to their own system. NFS and NIS are frequently used together: NIS makes sure that configuration information is propagated to all hosts, and NFS ensures that the files a user needs are accessible from these hosts.

NFS is also built on the RPC protocol and imposes a client-server relationship on the hosts that use it. An NFS server is a host that owns one or more filesystems and makes them available on the network; NFS clients mount filesystems from one or more servers. This follows the normal client-server model where the server owns a resource that is used by the client. In the case of NFS, the resource is a physical disk drive that is shared by all clients of the server.

There are two aspects to system administration using NFS: choosing a filesystem naming and mounting scheme, and then configuring the servers and clients to adhere to this scheme. The goal of any naming scheme should be to use network transparency wisely. Being able to mount filesystems from any server is useful only if the files are presented in a manner that is consistent with the users' expectations.

If NFS has been set up correctly, it should be transparent to the user. For example, if locally developed applications were found in */usr/local/bin* before NFS was installed, they should continue to be found there when NFS is running, whether */usr/local/bin* is on a local filesystem or a remote one. To the user, the actual disk holding */usr/local/bin* isn't important as long as the executables are accessible and built for the right machine architecture. If users must change their environments to locate files accessed through NFS, they will probably dislike the new network architecture because it changes the way things work.

An environment with many NFS servers and hundreds of clients can quickly become overwhelming in terms of management complexity. Successful system administration of a large NFS network requires adding some intelligence to the standard procedures. The cost of consistency on the network should not be a large administrative overhead. One tool that greatly eases the task of running an NFS network is the *automounter*, which applies NIS management to NFS configuration. This chapter starts with a quick look at how to get NFS up and running on clients and servers, and then explores NFS naming schemes and common filesystem planning problems. We'll cover the automounter in detail in

Setting up NFS

Setting up NFS on clients and servers involves starting the daemons that handle the NFS RPC protocol, starting additional daemons for auxiliary services such as file locking, and then simply exporting filesystems from the NFS servers and mounting them on the clients.

On an NFS client, you need to have the *lockd* and *statd* daemons running in order to use NFS. These daemons are generally started in a boot script (Solaris uses */etc/init.d/nfs.client*):

```
if [ -x /usr/lib/nfs/statd -a -x /usr/lib/nfs/lockd ]
then
    /usr/lib/nfs/statd > /dev/console 2>&1
    /usr/lib/nfs/lockd > /dev/console 2>&1
fi
```

On some non-Solaris systems, there may also be *biod* daemons that get started. The *biod* daemons perform block I/O operations for NFS clients, performing some simple read-ahead and write-behind performance optimizations. You run multiple instances of *biod* so that each client process can have multiple NFS requests outstanding at any time. Check your vendor's documentation for the proper invocation of the *biod* daemons. Solaris does not have *biod* daemons because the read-ahead and write-behind function is handled by a tunablr number of asynchronous I/O threads that reside in the system kernel.

The *lockd* and *statd* daemons handle file locking and lock recovery on the client. These locking daemons also run on an NFS server, and the client-side daemons coordinate file locking on the NFS server through their server-side counterparts. We'll come back to file locking later when we discuss how NFS handles state information.

On an NFS server, NFS services are started with the *nfsd* and *mountd* daemons, as well as the file locking daemons used on the client. You should see the NFS server daemons started in a boot script (Solaris uses */etc/init.d/nfs.server*):

```
if grep -s nfs /etc/dfs/sharetab >/dev/null ; then
    /usr/lib/nfs/mountd
    /usr/lib/nfs/nfsd -a 16
fi
```

On most NFS servers, there is a file that contains the list of filesystems the server will allow clients to mount via NFS. Many servers store this list in */etc/exports* file. Solaris stores the list in */etc/dfs/dfstab*. In the previous script file excerpt, the NFS server daemons are not started unless the host shares (exports) NFS filesystems in the */etc/dfs/dfstab* file. (The reference to */etc/dfs/sharetab* in the script excerpt is not a misprint; If there are filesystems to be made available for NFS service, the machine initializes the export list and starts the NFS daemons. As with the client-side, check your vendor's documentation or the boot scripts themselves for details on how the various server daemons are started.

The *nfsd* daemon accepts NFS RPC requests and executes them on the server. Some servers run multiple copies of the daemon so that they can handle several RPC requests at once. In Solaris, a single copy of the daemon is run, but multiple threads run in the kernel to provide parallel NFS service.

Network File System Design and Operation

It's possible to configure and use the Network File System without too much knowledge of how it is implemented or why various design decisions were made. But if you need to debug problems, or analyze patterns of NFS usage to suggest performance optimizations, you will need to know more about the inside workings of the NFS protocol and the daemons that implement it. With an understanding of how and why NFS does the things it does, you can more readily determine why it is broken or slow — probably the two most common complaints in any large NFS network.

Like NIS, NFS is implemented as a set of RPC procedures that use eXternal Data Representation (XDR) encoding to pass arguments between client and server. A filesystem mounted using NFS provides two levels of transparency:

- The filesystem appears to be resident on a disk attached to the local system, and all of the filesystem entries — files and directories — are viewed the same way, whether local or remote. NFS hides the location of the file on the network.
- NFS-mounted filesystems contain no information about the file server from which they are mounted. The NFS file server may be of a different architecture or running an entirely different operating system with a radically different filesystem structure. For example, a Sun machine running Solaris can mount an NFS filesystem from a Windows NT system or an IBM MVS mainframe, using NFS server implementations for each of these systems. NFS hides differences in the underlying remote filesystem structure and makes the remote filesystem appear to be of the exact same structure as that of the client.

NFS achieves the first level of transparency by defining a generic set of filesystem operations that are performed on a *Virtual File System* (VFS). The second level comes from the definition of *virtual nodes*, which are related to the more familiar Unix filesystem *inode* structures but hide the actual structure of the physical filesystem beneath them. The set of all procedures that can be performed on files is the vnode interface definition. The vnode and VFS specifications together define the NFS protocol.

Virtual Filesystems and Virtual Nodes

The Virtual File System allows a client system to access many different types of filesystems as if they were all attached locally. VFS hides the differences in implementations under a consistent interface. On a Unix NFS client, the VFS interface makes all NFS filesystems look like Unix filesystems, even if they are exported from IBM MVS or Windows NT servers. The VFS interface is really nothing more than a switchboard for filesystem- and file-oriented operations, as shown in Fig. 2

Actions that operate on entire filesystems, such as getting the amount of free space left in the filesystem, are called *VFS operations*; calls that operate on files or directories are *vnode operations*. On the server side, implementing a VFS entails taking the generic VFS and vnode operations and converting them into the appropriate actions on the real, underlying filesystem. This conversion happens invisibly to the NFS client process. It made a straightforward system call, which the client-side VFS turned into a vnode operation, and the server then converted into an equivalent operation on its filesystem.

For example, the *chown*( ) system call has an analogous operator in the vnode interface that sets the attributes of a file, as does the *stat*( ) system call that retrieves these attributes. There is not a strict one-to-one relationship of Unix system calls to vnode operations. The *write*( ) system call uses several filesystem calls to get a file's attributes, and append or modify blocks in the file. Some vnode operations are not defined on certain types of filesystems. The FAT filesystem, for example, doesn't have an equivalent of symbolic links, so an NFS file server running on an Windows NT machine rejects any attempts to use the vnode operation to create a symbolic link.

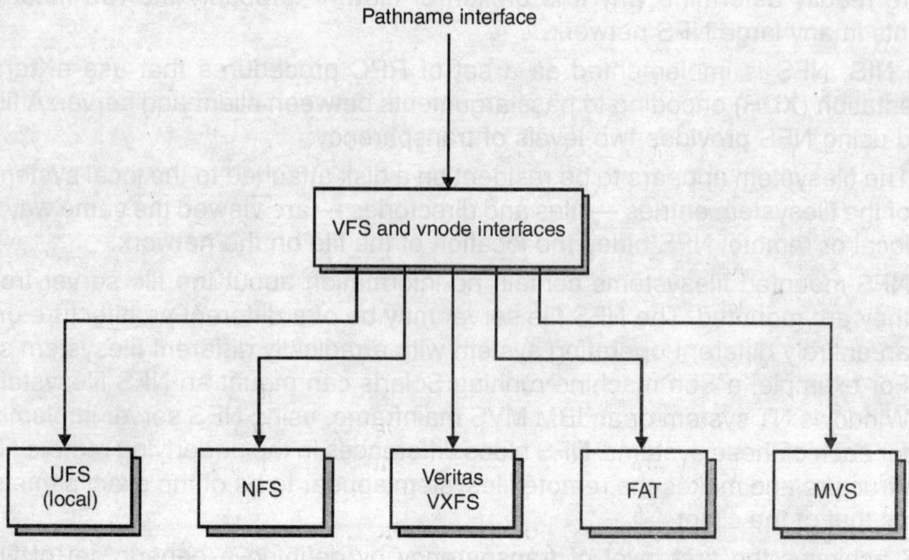

Fig. 2. Virtual File System interfaces

So far we have defined an interface to some filesystem objects, but not the mechanism used to "name" objects in the system. In a local Unix system call, these object names are file descriptors, which uniquely identify a file within the scope of a process. The counterparts of file descriptors in NFS are filehandle*s*, which are opaque "pointers" to files on the remote system. An opaque handle is of no value to the client because it can only be interpreted in the context of the remote filesystem. When you want to make a system call on a file, you first get a file descriptor for it. To make an NFS call (in the kernel) you must get a filehandle for the vnode. It is up to the virtual filesystem layer to translate user-level file descriptors into kernel-level filehandles.

Network Security

The simplicity and transparency provided by NFS and NIS must be weighed against security concerns. Providing access to all files to all users may not be in the best interests of security, particularly if the files contain sensitive or proprietary data. Not all hosts may be considered equally secure or "open," so access may be restricted to certain users. Transparency must be limited when dealing with secured hosts: if you have taken precautions to prevent unauthorized access to a machine, you don't want someone to be able to sit down and use an open window or logged-in terminal to access the secured machine. To enforce access restrictions, you always want password verification for users, which means eliminating some of the network transparency provided by NIS.

User-Oriented Network Security

One area of concern is user access to hosts on the network. Figure shows several classes of permissions to consider, reflecting the ways in which a user might access a host from another host on the network.

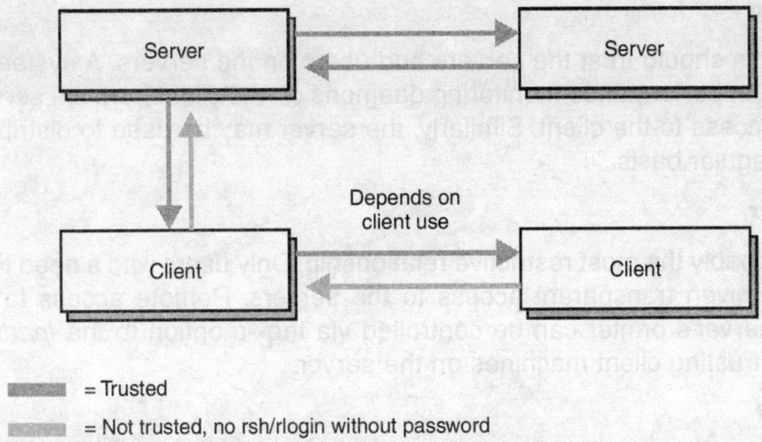

Fig. 3. Client-server remote logins

Remote logins are not the only concern; remote execution of commands using *rsh* should be considered in the same context. This section covers only login restrictions; we'll look at protecting data in NFS filesystems later in this chapter. Local login restrictions are defined by the local host's password file, NIS password maps, and the use of netgroups. Across the network, access is determined by the notion of trusted hosts and trusted users.

Trusted Hosts and Trusted Users

Defining a *trusted host* requires two machines: one that will be trusted and one that is extending the trust to it. The local host *lh* trusts remote host *rh* if users can log into *lh* from *rh* without supplying their passwords. Similarly, a user is trusted if he or she can log into a host from some remote machine without supplying a password. Trust is defined only for the local host; users and machines may be trusted on some systems but not on others.

The relationships between hosts often define the realm of trusted users and trusted hosts. Two NIS or NFS clients, for example, may trust all users and all other client hosts. On the NFS server, only other servers may be trusted hosts and only the system administration staff may be trusted users.

Warning : The following trusted user and trusted host descriptions apply in an environment in which you do not have to be wary of users or outsiders who will attempt to compromise security. These are basic security measures that fit in with the other network management strategies discussed in this book. If you need to secure your systems against all attacks, then you must consider the effects of having security compromised on any machine in your network. Again, these extensive security mechanisms are discussed in *Practical Unix Security*.

Some of the common patterns of trusting hosts and users are:

Server-Server

Generally, servers trust each other. A few users can be trusted in server-to-server relationships if each server has a password file that contains a subset of the NIS password map, or a password file with no NIS references. To emphasize the previous warning, extending trust between servers means that if one server is compromised, then they all are.

Server-Client

Most clients should trust the servers and users on the servers. A system administrator may need to run performance monitoring daemons on the client from the server and require transparent access to the client. Similarly, the server may be used to distribute files to the clients on a regular basis.

Client-Server

This is probably the most restrictive relationship. Only users with a need to use a service are generally given transparent access to the servers. Remote access to the server for access to a server's printer can be controlled via the *-u* option to the *lpadmin* command, instead of by trusting client machines on the server.

Client-Client

Client-client relationships depend upon how you have centralized your disk resources. If all files live on one or more fileservers, then client-to-client relationships are generally relaxed. However, if you are using the clients as isolated systems, with some per-client storage containing private data, then client-client relationships look more like those between clients and servers. The scope of the client-client relationships depends upon the sensitivity of the data on the clients: if you don't want other users to see the private data, then you must treat the client machine like a server.

The */etc/hosts.equiv* and *.rhosts* files (in each user's home directory) define the set of trusted hosts, users, and user-host pairs for each system. Again, trust and transparent access are granted by the machine being accessed remotely, so these configuration files vary from host to host. The *.rhosts* file is maintained by each user and specifies a list of hosts or user-host pairs that are also parsed for determining if a host or user is trusted.

Enabling Transparent Access

Both *rlogin* and *rsh* use the *ruserok( )* library routine to bypass the normal login and password security mechanism. The *ruserok( )* routine is invoked on the server side of a connection to see if the remote user gets transparent (i.e., no password prompt) access. To understand the semantics, let's look at its function prototype:

```
int ruserok(const char *rhost, int suser, const char *ruser,
        const char *luser);
```

The *rhost* parameter is the name of the remote host from where the remote user is. The *ruser* parameter is the login name of the remote user. The *luser* parameter is the name of local login name that the remote user wants transparent access to. Often *luser* and *ruser* are the same, but not always. The *suser* parameter is set to 1 if the UID of luser is 0, i.e., superuser. Otherwise, *suser* is set to 0.

ruserok*( ) checks first if* luser *exists; i.e., does* getpwnam*( ) return success for* luser *? It then determines if the remote user and hostname given are trusted on the* local *host; it is usually called by the remote daemon for these utilities during its startup. If the user or host are not trusted, then the user must supply a password to log in or get "Permission denied" errors when attempting to use* rsh. *If the remote host trusts the user and host, execution (or login) proceeds without any other verification of the user's identity.*

The *hosts.equiv* file contains either hostnames or host-user pairs:

```
hostname  [username]
```

If a username follows the hostname, only that combination of user and hostnames is trusted. Netgroup names, in the form *+@group*, may be substituted for either hostnames or usernames. As with the password file, using a plus sign (+) for an entry includes the appropriate NIS map: in the first column, the hosts map is included, and in the second column, the password map is included. Entries that grant permission contain the hostname, a host and username, or a netgroup inclusion.

The following is */etc/hosts.equiv* on host *mahimahi*:

```
wahoo
bitatron +
corvette johnc
+@source-hosts
+@sysadm-hosts  +@sysadm-users
```

The first example trusts all users on host *wahoo*. Users on *wahoo* can *rlogin* to *mahimahi* without a password, but only if the *ruser* and *luser* strings are equal.The second example is similar to the first, except that any remote user from *bitatron* can claim to be any local user and get access as the local user; i.e., *luser* and *ruser* do not have to be equal. This is certainly useful to the users who have access to *bitatron*, but it is very relaxed (or lax) security on *mahimahi*. The third example is the most restrictive. Only user *johnc* is trusted on host *corvette*, and of course *luser* and *ruser* (both "johnc") must be the same. Other users on host *corvette* are not trusted and must supply a password when logging in to *mahimahi*.

The last two entries use netgroups to define lists of hosts and users. The *+@source-hosts* entry trusts all hosts whose names appear in the *source-hosts* netgroup. If usernames are given as part of the netgroup triples, they are ignored. This means that hostname wildcards grant overly generous permissions. If the *source-hosts* netgroup contained *(,stern,)*, then using this netgroup in the first column of *hosts.equiv* effectively opens up the machine to all hosts on the network. If you need to restrict logins to specific users from specific machines, you must use either explicit names or netgroups in both the first and second column of *hosts.equiv*.

The last example does exactly this. Instead of trusting one host-username combination, it trusts all combinations of hostnames in *sysadm-hosts* and the usernames in *sysadm-users*. Note that the *usernames* in the *sysadm-hosts* netgroup and the *hostnames* in the *sysadm-users* netgroup are completely ignored.

Permission may be revoked by preceding the host or user specification with a minus sign (-):

```
-wahoo
+ -@dangerous-users
```

The first entry denies permission to all users on host *wahoo*. The second example negates all users in the netgroup *dangerous-users* regardless of what machine they originate from (the plus sign (+) makes the remote machine irrelevant in this entry).

If you want to deny permission to everything in both the hosts and password NIS maps, leave *hosts.equiv* empty.

The *.rhosts* file uses the same syntax as the *hosts.equiv* file, but it is parsed *after hosts.equiv*. The sole exception to this rule is when granting remote permission to *root*. When the superuser attempts to access a remote host, the *hosts.equiv* file is ignored and only the */.rhosts* file is read. For all other users, the *ruserok( )* routine first reads *hosts.equiv*. If it finds a positive match, then transparent access is granted. If it finds a negative match, and there is no *.rhosts* file for *luser*, then transparent access is denied. Otherwise, the *luser* 's *.rhosts* file is parsed until a match, either positive or negative, is found. If an entry in either file denies permission to a remote user, the file parsing stops at that point, even if an entry further down in the file grants permission to that user and host combination.

Usernames that are not the same on all systems are handled through the user's *.rhosts* file. If you are user *julie* on your desktop machine *vacation*, but have username *juliec* on host *starter*, you can still get to that remote host transparently by adding a line to your *.rhosts* file on *starter*. Assuming a standard home directory scheme, your *.rhosts* file would be */home/juliec/.rhosts* and should contain the name of the machine you are logging in from and your username on the originating machine:

```
vacation julie
```

From *vacation*, you can execute commands on *starter* using:

```
% rsh starter -l juliec "ls -l"
or:
% rlogin starter -l juliec
```

On *starter*, the *ruserok( )* routine looks for a *.rhosts* file for user *juliec*, your username on that system. If no entry in *hosts.equiv* grants you permission (probably the case because you have a different username on that system), then your *.rhosts* file entry maps your local username into its remote equivalent. You can also use netgroups in *.rhosts* files, with the same warnings that apply to using them in */etc/hosts.equiv*.

As a network manager, watch for overly permissive *.rhosts* files. Users may accidentally grant password-free access to any user on the network, or map a foreign username to their own Unix username. If you have many password files with private, non-NIS managed entries, watch the use of *.rhosts* files. Merging password files to eliminate non-uniform usernames may be easier than maintaining a constant lookout for unrestricted access granted through a *.rhosts* file.

Using Netgroups

Netgroups have been used in several examples already to show how triples of host, user, and domain names are used in granting access across the network. The best use of netgroups is for the definition of splinter groups of a large NIS domain, where creating a separate NIS domain would not justify the administrative effort required to keep the two domains synchronized.

Because of the variety of ways in which netgroups are applied, their use and administration are sometimes counterintuitive. Perhaps the most common mistake is defining a netgroup with host or usernames not present in the NIS maps or local host and password files. Consider a netgroup that includes a hostname in another NIS domain:

```
remote-hosts     (poi,-,-), (muban,-,-)
```

When a user attempts to *rlogin* from host *poi*, the local server-side daemon attempts to find the hostname corresponding to the IP address of the originating host. If *poi* cannot be found in the NIS *hosts.byaddr* map, then an IP address, instead of a hostname, is passed to *ruserok( )*. The verification process fails to match the hostname, even though it appears in the netgroup. Any time information is shared between NIS domains, the appropriate entries must appear in both NIS maps for the netgroup construction to function as expected.

Even though netgroups are specified as host and user pairs, no utility uses both names together. There is no difference between the following two netgroups:

```
group-a (los, mikel,) (bitatron, stern, )
group-b (los, -,) (bitatron, -,) (-, mikel, ) (-, stern, )
```

Things that need hostnames — the first column of *hosts.equiv* or NFS export lists — produce the set of hosts {*los, bitatron*} from both netgroups. Similarly, anything that takes a username, such as the password file or the second column of *hosts.equiv*, always finds the set {*mikel, stern*}. You can even mix-and-match these two groups in *hosts.equiv*. All four of the combinations of the two netgroups, when used in both columns of *hosts.equiv*, produce the same net effect: users *stern* and *mikel* are trusted on hosts *bitatron* and *los*.

The triple-based format of the netgroups map clouds the real function of the netgroups. Because all utilities parse either host or usernames, you will find it helpful to define netgroups that contain only host or usernames. It's easier to remember what each group is supposed to do, and the time required to administer a few extra netgroups will be more than made up by time not wasted chasing down strange permission problems that arise from the way the *netgroups* map is used.

An example here helps to show how the *netgroup* map can produce unexpected results. We'll build a netgroup containing a list of users and hosts that we trust on a server named *gate*. Users in the netgroup will be able to log in to *gate*, and hosts in the netgroup will be able to mount filesystems from it. The netgroup definition looks like this:

```
gate-group    (,stern,), (,johnc,), (bitatron, -,),
(corvette, -,)
```

In the */etc/dfs/dfstab* file on *gate*, we'll add a host access restriction:

share -o rw=gate-group /export/home/gate

No at-sign (@) is needed to include the netgroup name in the */etc/dfs/dfstab* file. The *netgroup* map is searched first for the names in the *rw=* list, followed by the *hosts* map.

In */etc/hosts.equiv* on *gate*, we'll include the *gate-group* netgroup:

```
+ +@gate-group
```

To test our access controls, we go to a machine not in the netgroup — NFS client *vacation* — and attempt to mount */export/home/gate*. We expect that the *mount* will fail with a "Permission denied" error:

```
vacation# mount gate:/home/gate/home/gate /mnt
vacation#
```

The mount completes without any errors. Why doesn't this netgroup work as expected?

The answer is in the wildcards left in the host fields in the netgroup entries for users *stern* and *johnc*. Because a wildcard was used in the host field of the netgroup, *all* hosts in the NIS map became part of *gate-group* and were added to the access list for */export/home/gate*. When creating this netgroup, our intention was probably to allow users *stern* and *johnc* to log in to *gate* from any host on the network, but instead we gave away access rights.

A better way to manage this problem is to define two netgroups, one for the users and one for the hosts, so that wildcards in one definition do not have strange effects on the other. The modified */etc/netgroup* file looks like this:

```
gate-users:    (,stern,), (,johnc,)
gate-hosts:    (bitatron,,), (corvette,,)
```

In the */etc/dfs/dfstab* file on *gate*, we use the *gate-hosts* netgroup:

```
share -o rw=gate-hosts /export/home/gate
```

and in */etc/hosts.equiv*, we use the netgroup *gate-users*. When host information is used, the *gate-hosts* group explicitly defines those hosts in the group; when usernames are needed, the *gate-users* map lists just those users. Even though there are wildcards in each group, those wildcards are in fields that are not referenced when the maps are used in these function-specific way

How secure are NIS and NFS?

NFS and NIS have bad reputations for security. NFS earned its reputation because of its default RPC security flavor AUTH_SYS is very weak. There are better security flavors available for NFS on Solaris and other systems. However, the better security flavors are not available for all, or even most NFS implementations, resulting in a practical dilemma for you. The stronger the NFS security you insist on, the more homogenous your computing environment will become. Assuming that secure file access across the network is a requirement, another option to consider is to not run NFS and switch to another file access system. Today there are but two practical choices:

SMB (also known as CIFS)

This limits your desktop environment to Windows. However, as discussed in, if you want strong security, you'll have to have systems capable of it, which means running Windows clients and servers throughout.

DCE/DFS

At the time this book was written, DCE/DFS was available as an add-on product developed by IBM's Pittsburgh Laboratory (also known as Transarc) unit for Solaris, IBM's AIX, and Windows. Other vendors offer DCE/DFS for their own operating systems (for example, HP offers DCE/DFS). So DCE/DFS offers the file access solution that is both heterogeneous and very secure.

NIS has earned its reputation because it has no authentication at all. The risk of this is that a successful attacker could provide a bogus NIS map to your users by having a host he controls masquerade as an NIS server. So the attacker could use a bogus host map to redirect the user to a host he controls (of course DNS has the same issue). Even more insidious, the attacker could gain root access when logging into a system, simply by providing a bogus passwd map. Another risk is that the encrypted password field from the *passwd* map in NIS is available to everyone, thus permitting attackers to perform faster password guessing than if they manually tried passwords via login attempts.

An enhancement to DNS, DNSSEC has been standardized but it is not widely deployed.

Ultimately, the future of directory services is LDAP, but at the time this book was written, the common security story for LDAP on Solaris, AIX, HP/UX, and Linux was not as strong as that of NIS+. You can get very secure LDAP out of Windows 2000, but then your clients and servers will be limited to running Windows 2000.

Password and NIS Security

Several volumes could be written about password aging, password guessing programs, and the usual poor choices made for passwords. Again, this book won't describe a complete password security strategy, but here are some common-sense guidelines for password security:

- Watch out for easily guessed passwords. Some obvious bad password choices are: your first name, your last name, your spouse or a sibling's name, the name of your favorite sport, and the kind of car you drive. Unfortunately, enforcing any sort of password approval requires modifying or replacing the standard NIS password management tools.
- Define and repeatedly stress local password requirements to the user community. This is a good first-line defense against someone guessing passwords, or using a password cracking program (a program that tries to guess user passwords using a long list of words). For example, you could state that all passwords had to contain at least six letters, one capital and one non-alphabetic character.
- Remind users that almost any word in the dictionary can be found by a thorough password cracker.
- Use any available password guessing programs that you find, such as Alec Muffet's *crack*. Having the same weapons as a potential intruder at least levels the playing field.

Managing the Root Password With NIS

NIS can be used to solve a common dilemma at sites with advanced, semi-trusted users. Many companies allow users of desktop machines to have the root password on their local hosts to install software, make small modifications, and power down/boot the system without the presence of a system administrator. With a different, user-specific root password on every system, the job of the system administrator quickly becomes a nightmare. Similarly, using the same root password on all systems defeats the purpose of having one.

Root privileges on servers should be guarded much more carefully, since too many hands touching host configurations inevitably creates untraceable problems. It is important to stress to semi-trusted users that their lack of root privileges on servers does not reflect a lack of expertise or trust, but merely a desire to exert full control over those machines for which you have full and total responsibility. Any change to a server that impacts the entire network becomes your immediate problem, so you should have jurisdiction over those hosts.

One way to discourage would-be part-time superusers is to require anyone with a server root password to carry the 24-hour emergency beeper at least part of each month.

Some approach is required that allows users to gain superuser access to their own hosts, but not to servers. At the same time, the system administrator must be able to become root on any system at any time to perform day-to-day maintenance. To solve the second problem, a common superuser password can be managed by NIS. Add an entry to the NIS password maps that has a UID of 0, but login name that is something other than *root*. For example, you might use a login name of *netroot*. Make sure the */etc/nsswitch.conf* file on each host lists *nis* on the *passwd:* entry:

```
passwd: files nis
```

Users are granted access to their own host via the *root* entry in the */etc/passwd* file.

Instead of creating an additional root user, some sites use a modified version of *su* that consults a "personal" password file. The additional password file has one entry for each user that is allowed to become root, and each user has a unique root password With either system, users are able to manage their own systems but will not know the root passwords on any other hosts. The NIS-managed *netroot* password ensures that the system administration staff can still gain superuser access to every host.

NFS security

Filesystem security has two aspects: controlling access to and operations on files, and limiting exposure of the contents of the files. Controlling access to remote files involves mapping Unix file operation semantics into the NFS system, so that certain operations are disallowed if the remote user fails to provide the proper credentials. To avoid giving superuser permissions across the network, additional constraints are put in place for access to files by *root*. Even more stringent NFS security requires proving that the Unix-style credentials contained in each NFS request are valid; that is, the server must know that the NFS client's request was made by a valid user and not an imposter on the network.

Limiting disclosure of data in a file is more difficult, as it usually involves encrypting the contents of the file. The client application may choose to enforce its own data encryption and store the file on the server in encrypted form. In this case, the client's NFS requests going over the network contain blocks of encrypted data. However, if the file is stored and used in clear text form, NFS requests to read or write the file will contain clear text as well. Sending parts of files over a network is subject to some data exposure concerns. In general, if security would be compromised by any part of a file being disclosed, then either the file should not be placed on an NFS-mounted filesystem, or you should use a security mechanism for RPC that encrypts NFS remote procedure calls and responses over the network. We will cover one such mechanism later in this section.

You can prevent damage to files by restricting write permissions and enforcing user authentication. With NFS you have the choice of deploying some simple security mechanisms and more complex, but stronger RPC security mechanisms..

RPC Security

Under the default RPC security mechanism, AUTH_SYS, every NFS request, including mount requests, contains a set of user credentials with a UID and a list of group IDs (GIDs) to which the UID belongs. NFS credentials are the same as those used for accessing local files, that is, if you belong to five groups, your NFS credentials contain your UID and five GIDs. On the NFS server, these credentials are used to perform the permission checks that are part of Unix file accesses — verifying write permission to remove a file, or execute permission to search directories. There are three areas in which NFS credentials may not match the user's local credential structure: the user is the superuser, the user is in too many groups, or no credentials were supplied (an "anonymous" request). Mapping of root and anonymous users is covered in the next section.

Problems with too many groups depend upon the implementation of NFS used by the client and the server, and may be an issue only if they are different (including different revisions of the same operating system). Every NFS implementation has a limit on the number of groups that can be passed in a credentials structure for an NFS RPC. This number usually agrees with the maximum number of groups to which a user may belong, but it may be smaller. On Solaris 8 the default and maximum number of groups is 16 and 32, respectively. However, under the AUTH_SYS RPC security mechanism, the maximum is 16. If the client's group limit is larger than the server's, and a user is in more groups than the server allows, then the server's attempt to parse and verify the credential structure will fail, yielding error messages like:

RPC: Authentication error

Authentication errors may occur when trying to mount a filesystem, in which case the superuser is in too many groups. Errors may also occur when a particular user tries to access files on the NFS server; these errors result from any NFS RPC operation. Pay particular attention to the *group* file in a heterogeneous environment, where the NIS-managed *group* map may be appended to a local file with several entries for common users like *root* and *bin*. The only solution is to restrict the number of groups to the smallest value allowed by all systems that are running NFS.

Stronger security for NFS

The security mechanisms described so far in this chapter are essentially refinements of the standard Unix login/password and file permission constraints, extended to handle distributed environments. Some additional care is taken to restrict superuser access over the network, but nothing in RPC's AUTH_SYS authentication protocol ensures that the user specified by the UID in the credential structure is permitted to use the RPC service, and nothing verifies that the user (or user running the application sending RPC requests) is really who the UID professes to be.

Simply checking user credentials is like giving out employee badges: the badge holder is given certain access rights. Someone who is not an employee could steal a badge and gain those same rights. Validating the user credentials in an NFS request is similar to making employees wear badges with their photographs on them: the badge grants certain access rights to its holder, and the photograph on the badge ensures that the badge holder is the "right" person. Stronger RPC security mechanisms than AUTH_SYS exist, which add credential validation to the standard RPC system. These stronger mechanisms can be used with NFS. We will discuss two of the stronger RPC security mechanisms available with Solaris 8, AUTH_DH, and RPCSEC_GSS. Both mechanisms rely on cryptographic techniques to achieve stronger security.

Security Services

Security isn't a monolithic concept, but among others, includes notions like authorization, auditing, and compartmentalization. RPC security is concerned with four services: identification, authentication, integrity, and privacy. Identification is merely the name RPC gives to the client and the server. The client's name usually corresponds to the UID. The server's name usually corresponds to the hostname. Authentication is the service that proves that the client and server are who they identify themselves to be. Integrity is the service that ensures the messages are not tampered with, or at least ensures that the receiver knows they have been tampered with. Privacy is the service that prevents eavesdropping.

Brief Introduction to Cryptography

Before we describe how the AUTH_DH and RPCSEC_GSS mechanisms work, we will explain some of the general principles of cryptography that apply to both mechanisms. A complete treatment of the topic can be found in the book *Applied Cryptography*, by Bruce Schneier (John Wiley and Sons, Inc., 1996).

There are four general cryptographic techniques that are pertinent:

- symmetric key encryption,
- asymmetric key encryption,
- public key exchange, and
- one way hash functions.

Symmetric Key Encryption

In a symmetric encryption scheme, the user knows some secret value (such as a password), which is used to encrypt a value such as a timestamp. The secret value is known as a *secret key*. The problem with symmetric encryption is that to get another host to validate your encrypted timestamp, you need to get your secret key (password) onto that host. Think of this problem as a password checking exercise: normally your password is verified on the local machine. If you were required to get your password validated on an NFS server, you or the system administrator would somehow have to get your password on that machine for it to perform the validation. An example of a symmetric key encryption scheme is the *Data Encryption Standard* (*DES*).

Asymmetric Key Encryption

Asymmetric key encryption involves the use of a public key to encrypt a secret value, such as a symmetric key, and, a private key to decrypt the same value. A public key and private key are associated as a pair. One half of the pair gets generated from the other via a series of arithmetic operations. The private key is never equal to the public key, hence the term *asymmetric*. As the names suggest, the public key is well-known to everyone, whereas the private key is known only to its owner. This helps solve the problem of getting a secret key on both hosts. You choose a symmetric secret key, encrypt it with the server's public key, send the result to the server and the server decrypts the secret key with its own private key. The secret key can then be used to encrypt a value like a timestamp, which the server validates by decrypting with the shared secret key. Alternatively, we could have encrypted the timestamp value with the server's public key, sent it to the server, and let the server decrypt it with the server's private key. However, asymmetric key encryption is usually much slower than symmetric key encryption. So, typically software that uses asymmetric key encryption uses symmetric key encryption once the shared secret key is established

The public key is published so that it is available for authentication services. The encryption mechanism used for asymmetric schemes typically uses a variety of exponentiation and other arithmetic operators that have nice commutative properties. The encryption algorithm is complex enough, and the keys themselves should be big enough (at least 1024 bits), to guarantee that a public key can't be decoded to discover its corresponding private key. Asymmetric key encryption is also called public key encryption. An example of an asymmetric key encryption is RSA.

Public Key Exchange

Public key exchange is similar to asymmetric key encryption in all ways but one: it does not encrypt a shared secret key with either public or private key. Instead, two agents, say a user and a server, generate a shared symmetric secret key that uniquely identifies one to the other but cannot be reproduced by a third agent, even if the initial agents' public keys are grabbed and analyzed by some attacker.

Here is how the shared secret key, also called a *common key*, is computed. The user sends to the server the user's public key, and the server sends to the user the server's public key. The user creates a common key by applying a set of arithmetic operations onto the server's public key and the user's private key.

The server generates the *same* key by applying the same arithmetic onto the user's public key and the server's private key. Because the algorithm uses commutative operations, the operation order does not matter — both schemes generate the same key, but *only* those two agents can recreate the key because it requires knowing at least one private key. An example of a public key exchange algorithm is Diffie-Hellman or DH for short.

One-way hash functions and MACs

A *one-way hash function* takes a string of octets of any length and produces a fixed width value called the *hash*. The function is designed such that given the hash, it is hard to find the string used as input to the one-way hash function, or for that matter, any string that produces the same hash result.

Let's say you and the server have established a common symmetric secret key using one of the three previously mentioned techniques. You now want to send a message to the server, but want to make sure an attacker in the middle cannot tamper with the message without the server knowing. What you can do is first combine your message with the secret key (you don't have to encrypt your message with the secret key), and then take this combination and apply the one way hash function to it. This computation is called a *message authentication code* or *MAC*. Then send both the MAC and the message (not the combination with the secret key) to the server. The server can then verify that you sent the message, and not someone who intercepted it by taking the message, combining it with the shared secret key in the same way you did, and computing the MAC. If the server's computed MAC is the same as the MAC you sent, the server has verified that you sent it.

Even though your message and MAC are sent in the clear to the server, an attacker in the middle cannot change the message without the server knowing it because this would change the result of the MAC computation on the server. The attacker can't change the MAC to match a tampered message because he doesn't know the secret key that only the server and you know. An example of a one-way hash function is MD5. An example of a MAC algorithm is HMAC-MD5.

When you add a MAC to a message you are enabling the security service of integrity.

NFS and IPSec

IPSec is the standard protocol for security at the IP network level. With IPSec you can beef up your trusted host relationships with strong cryptography. IPSec was invented by the Internet Engineering Task Force (IETF) to deal with three issues:

·Attackers are becoming quite adept at spoofing IP addresses. The attacker targets a host to victimize. The victim shares some resources (such as NFS exports) to only a specific set of clients and uses the source IP address of the client to check access rights. The attacker selects the IP address of one of these clients to masquerade as. Sometimes the attacker is lucky, and the client is down, so this is not too difficult. Or the attacker has to take some steps such as disabling a router or loading the targeted client. If the attacker fails, you might see messages like:

IP: Hardware address '%s' trying to be our address %s!

or:

IP: Proxy ARP problem? Hardware address '%s' thinks it is %s

on the legitimate client's console.

Once the legitimate client is disabled, the attacker then changes the IP address on a machine that he controls to that of the legitimate client and can then access the victim.

- An attacker that controls a gateway can easily engineer attacks where he tampers with the IP packets.
- Finally, if the Internet is to be a tool enabling more collaboration between organizations, then there needs to be a way to add privacy protections to sensitive traffic.

Here is what IPSec can do:

- Via per-host keys, allows hosts to authenticate each other. This frustrates IP spoofing attacks.
- Using a session key derived from per-host keys as input to a MAC, protects the integrity of IP traffic to frustrate packet tampering.
- Using a session key, encrypts all the data in the IP packet to frustrate eavesdropping.

The first two capabilities are provided by the *AH* (*Authentication Header*) feature of IPSec. The all three capabilities are provided by the *ESP* (*Encapsulating Security Payload*) feature of IPSec.

Many systems, including Solaris 8, have IPSec support. We won't go into the details of how to set up IPSec. However, we will point out that IPSec can be a useful tool to improve the security of your NFS environment:

- If you use the AH feature, then all NIS lookups are safe from tampering on the NIS server, or the NIS traffic. So the attacker cannot fool your NFS server into believing that client *gonzo* has IP address 192.4.5.6, instead of 10.1.2.3.
- Enabling AH on NFS clients and servers prevents attackers from spoofing the clients you list on servers' *rw=*, *ro=*, and *root=* lists.

➢ ESP used on the NFS client and server makes operating through firewalls safer when accessing sensitive data.

While IPSec is useful for securing NFS, because its security is host based, it does not protect your network from attackers that log onto your IPSec-protected hosts and assume the identity of other users. The discussions of AUTH_DH and RPCSEC_GSS that follow address this issue.

AUTH_DH: Diffie-Hellman Authentication

AUTH_DH is an RPC security flavor that uses encryption techniques to improve on AUTH_SYS.

Old terms: AUTH_DES, secure RPC, and, secure NFS

AUTH_DH was originally called AUTH_DES, and indeed, you'll find that most documentation of AUTH_DH still calls it AUTH_DES. However, the _DES part of the name is a misnomer. While the DES algorithm is used in AUTH_DH, as we will see later, it is the Diffie-Hellman algorithm that is central to the workings of the AUTH_DH. We will not refer to AUTH_DES again.

When AUTH_DH is combined with RPC, the combination is often referred to as "Secure RPC." When RPC, and AUTH_DH are combined with NFS, the result is often referred to as "Secure NFS." However, in the 13 years since AUTH_DH was invented, two things have occurred:

➢ AUTH_DH is no longer considered secure by many security experts. After you read the material on AUTH_DH, you might concur.

➢ RPCSEC_GSS, a stronger, more secure security flavor for RPC is now available with Solaris 8 and other NFS implementations.

We will not use the terms "Secure RPC" or "Secure NFS" again, since they are confusing. Instead, we will use RPC/DH and NFS/dh to refer to RPC secured with AUTH_DH, and NFS secured with RPC/DH, respectively.

Diffie-Hellman key exchange

AUTH_DH uses Diffie-Hellman public key exchange. Using this encryption scheme, RPC can be made more secure by requiring each client to establish a valid common key before making RPC requests to the server. Diffie-Hellman key exchange relies on each agent that wants to establish a common key to agree on two pieces of information beforehand. For AUTH_DH these pieces are:

➢ A base for the exponentiation part of the calculation. AUTH_DH uses a base of 3.

➢ A modulus used for the remainder part of the calculation. AUTH_DH uses a modulus of 0xd4a0ba0250b6fd2ec626e7efd637df76c716e22d0944b88b. Let's label this constant as AUTH_DH_MOD.

Let PRIV_C be the private key of the client. Then the public key of the client is:

(1) $PUBLIC\_C = (3^{PRIV\_C}) \bmod AUTH\_DH\_MOD$

Let PRIV_S be the private key of the server. Then the public key of the server is:

(2) $PUBLIC\_S = (3^{PRIV\_S}) \bmod AUTH\_DH\_MOD$

The client computes a common key between the client and server as:

(3) COMMON_C_S = (PUBLIC_S $^{PRIV\_C}$) mod AUTH_DH_MOD

The server computes the common key between the server and client as:

(4) COMMON_S_C = (PUBLIC_C $^{PRIV\_S}$) mod AUTH_DH_MOD

To prove that COMMON_C_S equals COMMON_S_C, we replace PUBLIC_S in statement (3) with the expression it was derived from in statement (2) and drop the MOD part of the expression. We do the same for PUBLIC_C in statement (4) with the expression from statement (2):

(3.1) COMMON_C_S_PRIME = $(3^{PRIV\_S})^{PRIV\_C} = 3^{PRIV\_S * PRIV\_C}$

(4.1) COMMON_S_C_PRIME = $(3^{PRIV\_C})^{PRIV\_S} = 3^{PRIV\_C * PRIV\_S}$

Because multiplication is a commutative operation, it is obvious that COMMON_C_S_PRIME equals COMMON_S_C_PRIME. Therefore, COMMON_C_S equals COMMON_S_C.

How RPC/DH works

RPC/DH uses a combination of Diffie-Hellman key exchange and DES encryption. User validation is performed by the server, based on information in the RPC request.

The common key will be used to construct a shared secret DES key. Note that because AUTH_DH_MOD is 192 bits, the common key will be 192 bits. However, DES uses 64 bit keys, such that the low order bit of each octet is a parity bit, making DES effectively a 56-bit symmetric key algorithm. AUTH_DH deals with this by selecting the middle 64 bits of the common key. These 64 bits are split into eight octets. Parity is added to the low order bit of each octet. In addition the high order bit of each octet is unused, making this effectively a 48-bit shared secret key.

The first time the client contacts the server, it generates a random session key, and encrypts it with the shared secret DES key. The session key is also a DES key. The client also generates a time-to-live value (in seconds) called the *window*, and a window value that is one second less than the first window value. The two window values are encrypted with the session key. The encrypted session key and the encrypted window values are sent to the server. The server can decrypt the encrypted session key because it knows the common key, and therefore the shared secret DES key. With the session key, it can decrypt the window values. If the second window value is not one less than the first, the server knows something odd is going on, and it rejects the client's request.

The first time, and on every subsequent contact to the server, the client encrypts the current time using the session key. It sends its RPC request to the server.

The server decrypts the timestamp, using the same session key, and verifies that it is accurate. If the decrypted timestamp falls outside of the time to live window, the server rejects the request.

So far we've described how RPC/DH does authentication. We will now look at how identity works in RPC/DH. Recall that AUTH_SYS sends a UID, GID, and a list of supplementary GIDs. The first time RPC/DH contacts the server to establish the session key, it sends no UIDs or GIDs. Instead it sends a string, called a *netname*, which identifies two items:

- The user (albeit, the username is a UID expressed in ASCII-decimal)
- The domain name of the user (usually this is an NIS domain name)

The server does three things with the netname:

- Locates an NIS server serving the specified domain that knows about the user.
- Looks up the user's netname in the NIS *netid* map for the user's UID, GID, and list of supplementary groups.
- Looks up the user's netname in the NIS *publickey* map for the user's Diffie-Hellman public key. With that, and the server's private key, the server can determine the common key, then the shared secret key, then decrypt the session key, and use that to verify that the request came from a user corresponding to the netname.

RPC/DH State and NFS Statelessness

The title of the section says it all. How can we reconcile the fact that NFS is stateless, and yet RPC/DH clearly establishes state in the form of a session key, with a time to live? This state has to be kept on the server. The answer is that this is not state that has to be recovered in the event of a server crash, which is in stark contrast to file locking state. If the server reboots, or if it even decides to throw away an RPC/DH session, it is not a disaster. The client simply gets an error indicating that the server has no knowledge of the session, and the client establishes a new session key as if it was the first contact between the client and server.

❑❑❑

Appendix B

Viruses

A computer virus is a piece of code that modifies the operating system or system utilities with harmful or annoying side effects. Like human viruses, a computer virus reproduces itself and spreads through a vector, or carrier. Once one computer is infected, the virus attempts to copy itself onto floppies or other removable media that will be taken to other systems. When an infected disk is inserted into a healthy system, the virus loads itself into the uninfected system. Entire networks of computers may be infected from a single disk that infects a system that later infects a file server, for example.

Effects of viruses vary greatly. Some simply render the machine useless, echoing annoying messages back to the user but preventing any "real" command execution. Others are destructive in nature, scribbling on critical filesystem information on hard disks or removing key files.

Viruses are virtually unknown in time-sharing operating systems such as Unix that enforce kernel protection. The operating system cannot be modified without superuser permission, so random user applications cannot inject viruses into the system. The Windows operating system, on the other hand, does not protect its kernel code or disk files, so an executable can overwrite parts of the kernel, the DOS image on disk, or various system utilities. Once the disk image is infected, the system remains infected, even through reboots or power cycles. Note that viruses are not the same as worms, rabbits, or other user-level processes that consume resources or reproduce rapidly enough to bring a system to a halt. A computer virus specifically damages the operating system.

Enforcing basic security around the root password and superuser access to machines should be sufficient to deter deliberate planting of viruses in the Unix kernel. In addition to securing access from the local area network, verify that your systems are safe from attacks from external networks such as the Internet. If you can prevent unauthorized superuser access, then you must only worry about things that you or your system administrators do as root.

Watch what you put into *cron* entries. Any script that gets run by cron should be owned by root and either not writable or writable only by root. If a user asks for a shell script to be added to root 's *crontab*, install the script so that the user cannot modify it once it has been added to the *crontab* file.

Similarly, avoid any package that requires an executable to be run as root as part of its installation process, unless you can vouch for the integrity of the package's provider. In general, vendors stand behind the safety of their software, and you should not worry about "branding" utilities that write serial number information into executable images or packages. It is becoming the norm for vendors to include a strong cryptographic checksum like MD5, or a verifiable digital signature with software on removable media or web sites. Such practices are a good sign that your vendor is taking care to secure its software from viruses, but such practices do you the most good when you take the time to verify the checksum or signature before installing the software.

The same guidelines that apply to Windows users also apply to Unix system administrators: if you don't know where an executable came from, don't run it as *root*. This is especially true for executables taken from public domain sources. If you can't get the source code, don't experiment with it unless you are willing to perform a post-installation check for damage. Above all else, use common sense. If you feel uncomfortable loaning your car keys to a complete stranger, you should feel equally queasy about installing strange software on your system as root.

Appendix C
Remote Procedure Call Tools

Network failures on a grand scale are generally caused by problems at the MAC or IP level, and are immediately noticed by users. Problems involving higher layers of the network protocol stack manifest themselves in more subtle ways, affecting only a few machines or particular pairs of clients and servers. The utilities discussed in the following sections analyze functionality from the remote procedure call (RPC) layer up through the NFS or NIS application layer. The next section contains a detailed examination of the RPC mechanism at the heart of NFS and NIS.

RPC Mechanics

The Remote Procedure Call (RPC) mechanism imposes a client/server relationship on machines in a network. A server is a host that physically owns some shared resource, such as a disk exported for NFS service or an NIS map. Clients operate on resources owned by servers by making RPC requests; these operations appear (to the client) to have been executed locally. For example, when performing a read RPC on an NFS-mounted disk, the reading application has no knowledge of where the read is actually executed. Many client-server relationships may be defined for each machine on a network; a server for one resource is often a client for many others in the same network.

Identifying RPC Services

Services available through RPC are identified by four values:

- Program number
- Version number
- Procedure number
- Protocol (UDP or TCP)

The program number uniquely identifies the RPC service. Each RPC service, such as the *mountd* or NIS server daemons, is assigned a program number. The file */etc/rpc* and the *rpc* NIS map contain an enumeration of RPC program numbers, formal names, and nicknames for each service:

Excerpt from /etc/rpc:

```
nfs          100003   nfsprog
ypserv       100004   ypprog
mountd       100005   mount showmount
ypbind       100007
```

Note that program 100005, *mountd*, has two names, reflecting the fact that the *mountd* daemon services both *mount* requests and the *showmount* utility.

Program numbers can also be expressed in hexadecimal. Well-known RPC services such as NFS and NIS are assigned reserved program numbers in the range 0x0 to 0x199999. Numbers above this range may be assigned to local applications such as license servers. The well-known programs are commonly expressed in decimal, though.

A version number is used to differentiate between various flavors of the same service, and is mostly utilized to evolve the service over time, while providing backwards compatibility if so desired. For example, there are two versions of the NFS service: Versions 2 and 3 (there is no Version 1). Each version of the program may be composed of many procedures. Each version of the NFS service, program number 100003, consists of several procedures, each of which is assigned a procedure number. These procedures perform client requests on the NFS server. For example: read a directory, create a file, read a block from a file, write to a file, get the file's attributes, or get statistics about a filesystem. The procedure number is passed in an RPC request as an "op code" for the RPC server. Procedure numbers start with 1; procedure 0 is reserved for a "null" function. While RPC program numbers are well-advertised, version and procedure numbers are particular to the service and often are contained in a header file that gets compiled into the client program. NFS procedure numbers, for example, are defined in the header files */usr/include/nfs/nfs.h*.

RPC clients and servers deal exclusively with RPC program numbers. At the session layer in the protocol stack, the code doesn't really care what protocols are used to provide the session services. The UDP and TCP transport protocols need port numbers to identify the local and remote ends of a connection. The portmapper is used to perform translation between the RPC program number-based view of the world and the TCP/UDP port numbers.

RPC Portmapper – rpcbind

The *rpcbind* daemon (also known as the portmapper), exists to register RPC services and to provide their IP port numbers when given an RPC program number. *rpcbind* itself is an RPC service, but it resides at a well-known IP port (port 111) so that it may be contacted directly by remote hosts. For example, if host *fred* needs to mount a filesystem from host *barney*, it must send an RPC request to the *mountd* daemon on *barney*. The mechanics of making the RPC request are as follows:

The *rpcbind* daemon and the old portmapper provide the same RPC service. The portmapper implements Version 2 of the portmap protocol (RPC program number 100000), where the *rpcbind* daemon implements Versions 3 and 4 of the protocol, in addition to Version 2. This means that the *rpcbind* daemon already implements the functionality provided by the old portmapper. Due to this overlap in functionality and to add to the confusion, many people refer to the *rpcbind* daemon as the portmapper.

- fred *gets the IP address for* barney*, using the* ipnodes *NIS map.* fred *also looks up the RPC program number for* mountd *in the* rpc *NIS map. The RPC program number for* mountd *is 100005.*
- Knowing that the portmapper lives at port 111, *fred* sends an RPC request to the portmapper on *barney*, asking for the IP port (on *barney*) of RPC program 100005. *fred* also specifies the particular protocol and version number for the RPC service. *barney* 's portmapper responds to the request with port 704, the IP port at which *mountd* is listening for incoming mount RPC requests over the specified protocol. Note that it is possible for the portmapper to return an error, if the specified program does not exist or if it hasn't been registered on the remote host. *barney*, for example, might not be an NFS server and would therefore have no reason to run the *mountd* daemon.
- fred *sends a* mount *RPC request to* barney*, using the IP port number returned by the portmapper. This RPC request contains an RPC procedure number, which tells the* mountd *daemon what to do with the request. The RPC request also contains the parameters for the procedure, in this case, the name of the filesystem* fred *needs to mount.*

The portmapper is also used to handle an *RPC broadcast*. Recall that a network broadcast is a packet that is sent to all hosts on the network; an RPC broadcast is a request that is sent to all servers for a particular RPC service. For example, the NIS client *ypbind* daemon uses an RPC broadcast to locate an NIS server for its domain. There's one small problem with RPC broadcasts: to send a broadcast packet, a host must fill in the remote port number, so all hosts receiving the packet know where to deliver the broadcast packet. RPC doesn't have any knowledge of port numbers, and the RPC server daemons on some hosts may be registered at different port numbers. This problem is resolved by sending RPC broadcasts to the portmapper, and asking the portmapper to make the RPC call indirectly on behalf of the sender. In the case of the *ypbind* daemon, it sends a broadcast to all *rpcbind* daemons; they in turn call the *ypserv* RPC server on each host.

RPC Version Numbers

As mentioned before, each new implementation of an RPC server has its own version number. Different version numbers are used to coordinate multiple implementations of the same service, each of which may have a different interface. As an RPC service matures, the service's author may find it necessary to add new procedures or add arguments to existing procedures. Changing the interface in this way requires incrementing the version number. The first (and earliest) version of an RPC program is version 1; subsequent releases of the server should use consecutive version numbers. For example, the mount service has several versions, each one supporting more options than its predecessors.

Multiple versions are implemented in a single server process; there doesn't need to be a separate instance of the RPC server daemon for each version supported. Each RPC server daemon registers its RPC program number and all versions it supports with the portmapper. It is helpful to think of dispatching a request through an RPC server as a two-level switch: the first level discriminates on the version number, and chooses a set of procedure routines comprising that version of the RPC service. The second level dispatch invokes one of the routines in that set based on the program number in the RPC request.

When contacting the portmapper on a remote host, the local and remote sides must agree on the version number of the RPC service that will be used. The rule of thumb is to use the highest-numbered version that both parties understand. In cases where version numbers are not consecutively numbered, or no mutually agreeable version number can be found, the portmapper returns a *version mismatch* error looking like:

```
mount: RPC: Program version mismatch
```

Even though Solaris supports Transport-Independent RPC (TI-RPC), in reality most RPC services use the TCP, UDP and loopback transport protocols. Servers may register themselves for any of the protocols, depending upon the varieties of connections they need to support. UDP packets are unreliable and unsequenced and are often used for broadcast or stateless services. The RPC server for the *spray* utility, which "catches" packets thrown at the remote host, uses the UDP protocol to accept as many requests as it can without requiring retransmission of any missed packets. In contrast to UDP, TCP packets are reliably delivered and are presented in the order in which they were transmitted, making them a requirement when requests must be processed by the server in the order in which they were transmitted by the client. The loopback transports are used for communication within the local host and can be connection-less or connection-oriented. For example, the automounter daemon uses RPC over a connection-oriented loopback transport to communicate with the local kernel.

RPC servers listen on the ports they have registered with the portmapper, and are used repeatedly for short-lived sessions. Connections to an RPC server may exist for the duration of the RPC call only, or may remain across calls. They do not usually fork new processes for each request, since the overhead of doing so would significantly impair the performance of RPC-intensive services such as NFS. Many RPC servers are multithreaded, such as NFS in Solaris, which allows the server to have multiple NFS requests being processed in parallel. A multithreaded NFS server can take advantage of multiple disks and disk controllers, it also allows "fast" NFS requests such as attribute or name lookups to not get trapped behind slower disk requests.

RPC Registration

Making RPC calls is a reasonably complex affair because there are several places for the procedure to break down. The *rpcinfo* utility is an analog of *ping* that queries RPC servers and their registration with the portmapper. Like *ping*, *rpcinfo* provides a measure of basic connectivity, albeit at the session layer in the network protocol stack. Pinging a remote machine ensures that the underlying physical network and IP address handling are correct; using *rpcinfo* to perform a similar test verifies that the remote machine is capable of accepting and replying to an RPC request.

rpcinfo *can be used to detect and debug a variety of failures:*

- "Dead" or hung servers caused by improper configuration or a failed daemon
- RPC program version number mismatches between client and server

- Bogus or renegade RPC servers, such as an NIS server that does not have valid maps for the domain it pretends to serve
- Broadcast-related problems

In its simplest usage, *rpcinfo -p* takes a remote hostname (or uses the local hostname if none is specified) and queries the portmapper on that host for all registered RPC services:

```
% rpcinfo -p corvette
      program vers proto   port  service
       100000    4   tcp    111  portmapper
       100000    3   tcp    111  portmapper
       100000    2   tcp    111  portmapper
       100000    4   udp    111  portmapper
       100000    3   udp    111  portmapper
       100000    2   udp    111  portmapper
       100024    1   udp  32781  status
       100024    1   tcp  32775  status
       100011    1   udp  32787  rquotad
       100002    2   udp  32789  rusersd
       100002    3   udp  32789  rusersd
       100002    2   tcp  32777  rusersd
       100002    3   tcp  32777  rusersd
       100021    1   udp   4045  nlockmgr
       100021    2   udp   4045  nlockmgr
       100021    3   udp   4045  nlockmgr
       100021    4   udp   4045  nlockmgr
       100021    1   tcp   4045  nlockmgr
       100021    2   tcp   4045  nlockmgr
       100021    3   tcp   4045  nlockmgr
       100021    4   tcp   4045  nlockmgr
       100012    1   udp  32791  sprayd
       100008    1   udp  32793  walld
       100001    2   udp  32795  rstatd
       100001    3   udp  32795  rstatd
       100001    4   udp  32795  rstatd
       100068    2   udp  32796  cmsd
       100068    3   udp  32796  cmsd
       100068    4   udp  32796  cmsd
       100068    5   udp  32796  cmsd
       100005    1   udp  32810  mountd
```

```
100005    2   udp   32810   mountd
100005    3   udp   32810   mountd
100005    1   tcp   32795   mountd
100005    2   tcp   32795   mountd
100005    3   tcp   32795   mountd
100003    2   udp    2049   nfs
100003    3   udp    2049   nfs
100227    2   udp    2049
100227    3   udp    2049
100003    2   tcp    2049   nfs
100003    3   tcp    2049   nfs
100227    2   tcp    2049
100227    3   tcp    2049
```

The output from *rpcinfo* shows the RPC program and version numbers, the protocols supported, the IP port used by the RPC server, and the name of the RPC service. Service names come from the *rpc.bynumber* NIS map; if no name is printed next to the registration information then the RPC program number does not appear in the map. This may be expected for third-party packages that run RPC server daemons, since the hardware vendor creating the */etc/rpc* file doesn't necessarily list all of the software vendors' RPC numbers. However, a well-known RPC service should be listed properly. Missing RPC service names could indicate a corrupted or incomplete *rpc.bynumber* NIS map. One exception is the NFS ACL service, defined as RPC program 100227. Solaris does not list it in */etc/rpc*, and therefore its name is not printed in the previous output. The NFS ACL service implements the protocol used between Solaris hosts to exchange ACL (Access Control List) information, though it is currently only interoperable between Solaris hosts. If the client or server do not implement the service, then traditional Unix file access control based on permission bits is used.

If the portmapper on the remote machine has died or is not accepting connections for some reason, *rpcinfo* times out attempting to reach it and reports the error. This is a good first step toward diagnosing any RPC-related problem: verify that the remote portmapper is alive and returning valid RPC service registrations.

rpcinfo *can also be used like* ping *for a particular RPC server:*

```
rpcinfo -u host program version          UDP-based services
rpcinfo -t host program version          TCP-based services
```

The *-u* or *-t* parameter specifies the transport protocol to be used — UDP or TCP, respectively. The hostname must be specified, even if the local host is being queried. Finally, the RPC program and version number are given; the program may be supplied by name (one reported by *rpcinfo -p*) or by explicit numerical value.

As a practical example, consider trying to mount an NFS filesystem from server *mahimahi*. You can mount it successfully, but attempts to operate on its files hang the client. You can use *rpcinfo* to check on the status of the NFS RPC daemons on *mahimahi*:

```
% rpcinfo -u mahimahi nfs 2
```

program 100003 version 2 ready and waiting

In this example, the NFS v2 RPC service is queried on remote host *mahimahi*. Since the service is specified by name, *rpcinfo* looks it up in the *rpc* NIS map. The *-u* flag tells *rpcinfo* to use the UDP protocol. If the *-t* option had been specified instead, *rpcinfo* would have reported the status of the NFS over TCP service. At the time of this writing, a handful of vendors still do not support NFS over TCP, therefore a *-t* query to one of their servers would report that *rpcinfo* could not find a registration for the service using such a protocol.

rpcinfo -u *and* rpcinfo -t *call the null procedure (procedure 0) of the RPC server. The null procedure normally does nothing more than return a zero-length reply. If you cannot contact the null procedure of a server, then the health of the server daemon process is suspect. If the daemon never started running,* rpcinfo *would have reported that it couldn't find the server daemon at all. If* rpcinfo *finds the RPC server daemon but can't get a null procedure reply from it, then the daemon is probably hung.*

Debugging RPC Problems

If the RPC service is not registered, or if you can't reach the RPC server daemon, it's likely there is a low-level problem in the network. However, sometimes you reach an RPC server, but you find the wrong one or it gives you the wrong answer. If you have a heterogeneous environment and are running multiple versions of each RPC service, it's possible to get RPC version number mismatch errors.

These problems affect NIS and diskless client booting; they are best sorted out by using *rpcinfo* to emulate an RPC call and by observing server responses. Networks with multiple, heterogeneous servers may produce multiple, conflicting responses to the same broadcast request. Debugging problems that arise from this behavior often require knowing the order in which the responses are received.

Here's an example: we'll perform a broadcast and then watch the order in which responses are received. When a diskless client boots, it may receive several replies to a request for boot parameters. The boot fails if the first reply contains incorrect or invalid boot parameter information. *rpcinfo -b* sends a broadcast request to the specified RPC program and version number. The RPC program can either be specified in numeric (100026) form, or in its name equivalent (bootparam):

```
% rpcinfo -b bootparam 1
      fe80::a00:20ff:feb5:1fba.128.67          unknown
      fe80::a00:20ff:feb9:2ad1.128.78          unknown
      131.40.52.238.128.67                     mora
      131.40.52.81.128.68                      kanawha
      131.40.52.221.128.79                     holydev
      Next Broadcast
% rpcinfo -b bootparam 1
      131.40.52.81.128.68                      kanawha
      fe80::a00:20ff:feb5:1fba.128.67          unknown
```

```
131.40.52.238.128.67                mora
fe80::a00:20ff:feb9:2ad1.128.78     unknown
131.40.52.221.128.79                holydev
Next Broadcast
```

In this example, a broadcast packet is sent to the boot parameter server (bootparam). *rpcinfo* obtains the RPC program number (100026) from */etc/rpc* or the *rpc.bynumber* NIS map (depending on */etc/nsswitch.conf*). Any host that is running the boot parameter server replies to the broadcast with the standard null procedure "empty" reply. The *universal address* for the RPC service is printed by the requesting host in the order in which replies are received from these hosts (see the sidebar). After a short interval, another broadcast is sent.

Universal Addresses

A universal address identifies the location of a transport endpoint. For UDP and TCP, it is composed of the dotted IP address with the port number of the service appended. In this example, the host *kanawha* has a universal address of 131.40.52.81.128.68.

The first four elements in the dotted string form the IP address of the server *kanawha*:

```
% ypmatch 131.40.52.81
hosts.byaddr
131.40.52.81    kanawha
```

The last two elements, "128.68", are the high and low octets of the port on which the service is registered (32836). This number is obtained by multiplying the high octet value by 2^8 and adding it to the low octet value:

```
128 * 2^8 = 32768      (high
octet)
+                68     (low octet)
---
32836    (decimal representation of port)
```

rpcinfo *helps us verify that* bootparam *is indeed registered on port 32836:*

```
% rpcinfo -p kanawha | grep
bootparam
100026 1 udp 32836 bootparam
```

Server loading may cause the order of replies between successive broadcasts to vary significantly. A busy server takes longer to schedule the RPC server and process the request. Differing reply sequences from RPC servers are not themselves indicative of a problem, if the servers all return the correct information. If one or more servers has incorrect information, though, you will see irregular failures. A machine returning correct information may not always be the first to deliver a response to a client broadcast, so sometimes the client gets the wrong response.

In the last example (diskless client booting), a client that gets the wrong response won't boot. The boot failures may be very intermittent due to variations in server loading: when the server returning an invalid reply is heavily loaded, the client will boot without problem.

However, when the servers with the correct information are loaded, then the client gets an invalid set of boot parameters and cannot start booting a kernel.

Binding to the wrong NIS server causes another kind of problem. A renegade NIS server may be the first to answer a *ypbind* broadcast for NIS service, and its lack of information about the domain makes the client machine unusable. Sometimes, just looking at the list of servers that respond to a request may flag a problem, if you notice that one of the servers should not be answering the broadcast:

```
% rpcinfo -b ypserv 1
131.40.52.138.3.255        poi
131.40.52.27.3.166         onaga
131.40.52.28.3.163         mahimahi
```

In this example, all NIS servers on the local network answer the *rpcinfo* broadcast request to the null procedure of the *ypserv* daemon. If *poi* should not be an NIS server, then the network will be prone to periods of intermittent failure if clients bind to it. Failure to fully decommission a host as an NIS server — leaving empty NIS map directories, for example — may cause this problem.

There's another possibility for NIS failure that *rpcinfo* cannot detect: there may be NIS servers on the network, but no servers for the client's NIS domain. In the previous example, *poi* may be a valid NIS server in another domain, in which case it is operating properly by responding to the *rpcinfo* broadcast. You might not be able to get *ypbind* started on an NIS client because all of the servers are in the wrong domain, and therefore the client's broadcasts are not answered. The *rpcinfo -b* test is a little misleading because it doesn't ask the NIS RPC daemons what domains they are serving, although the client's requests will be domain-specific. Check the servers that reply to an *rpcinfo -b* and ensure that they serve the NIS domain used by the clients experiencing NIS failures.

If a client cannot find an NIS server, *ypbind* hangs the boot sequence with errors of the form:

WARNING : Timed out waiting for NIS to come up.

Using *rpcinfo* as shown helps to determine why a particular client cannot start the NIS service: if no host replies to the *rpcinfo* request, then the broadcast packet is failing to reach any NIS servers. If the NIS domain name and the broadcast address are correct, then it may be necessary to override the broadcast-based search and hand *ypbind* the name and address of a valid NIS server

❑❑❑

PREVIOUS YEARS EXAMINATION PAPERS

B.E.

VIth SEMESTER EXAMINATION MAY 2006

NETWORK PROGRAMING (IT-302-E)

Time : 3 Hours *Total Marks : 100*

Note : Attempt any five questions.

1. (a) Explain the term Networking, what are its advantages.
   (b) Explain TCP/IP layer architecture.
2. Explain in detail TCP and UDP datagrams.
3. (a) Explain TCP Client-Server model.
   (b) What is socket and its structure ?
4. Explain connectionless and connection-oriented servers. Discuss various issues in server software design.
5. Explain remote procedure call concept. What is remote program and its multiple versions.
6. (a) What is N/W file system concept of datalink ?
   (b) What are various debugging techniques ?
7. (a) Explain Routing sockets.
   (b) How boradcasting is done to mobile network ?
8. Write short notes on :
   (i) Multi-service servers (TCP, UPD)
   (ii) Dynamic port mapping

B.E.

VIth SEMESTER EXAMINATION DEC. 2006

NETWORK PROGRAMING (IT-302-E)

Time : 3 Hours *Total Marks : 100*

Note : Attempt any five questions.

1. (a) Explain the architecture of TCP/IP protocol.
(b) What is ICMP and what is its purpose ?

2. Explain address resolution protocol and RARP.

3. (a) Explain the basic Unix I/O function used in socket programming, use suitable examples.
(b) What are various system calls used for sockets ?

4. Explain in brief various types of servers. What is Client-Server Model ?

5. (a) Explain multiprotocol server.
(b) Explain multi-service server.

6. What are remote procedure call models ? How they work ?

7. Explain network file system concept of data link access. What are debugging techniques ?

8. Write short notes on :
(i) Classful internet addresses
(ii) Super netting.

B.E.

VIth SEMESTER EXAMINATION DEC. 2007

NETWORK PROGRAMING (IT-302-E)

Time : 3 Hours *Total Marks : 100*

Note : Attempt any five questions. All question carry equal marks.

1. Explain the following commands with their format of usage and corresponding outputs :

(i) FINGER

(ii) NET Stat

(iii) IPCONFIG

(iv) Ping

2. (a) Explain classful addressing with its weakness.

(b) Explain the datagram format of IPV6.

3. Describe with complete signature, all the input function for elementary UDP client/ server.

4. What is an echo server ? Write a program to implement TCP echo server.

5. (a) Discuss the RPC model

(b) Explain Dynamic Port Mapping Algorithm.

6. (a) Discuss various NFS file modes with file attributes.

(b) What do you mean by debugging ? Explain various debugging techniques.

7. (a) Differentiate between interative and concurrent servers, discuss in reference to their areas of application.

(b) Discuss with process structure, the algorithm for concurrent connection oriented server.

8. Write short notes on :

(i) I/O Multiplexing

(ii) Multiprotocol and Multiservice servers.

(iii) Routing sockets

B.E.

VIth SEMESTER EXAMINATION MAY 2008

NETWORK PROGRAMING (IT-302-E)

Time : 3 Hours *Total Marks : 100*

Note : Attempt any five questions. All question carry equal marks.

1. (a) Explain TCP/IP protocol Architecture.

(b) Explain ICMP with its purpose.

(c) List out various advantages and disadvantages of computer networking.

2. Explain the following in context of socket (TCP and UDP)

(a) Elementary node

(b) Address Conversions

(c) Echo services

3. (a) Explain various debugging technique with examples.

(b) List out various difference between TCP and UDP datagrams.

4. Explain the following :

(a) RPC Transmission

(b) Dynamic Port Mapping

(c) Authentication in RPC

5. Explain the following :

(a) Multi service servers for TCP and UDP.

(b) Concurrent serves for TCP and UDP.

6. (a) Explain connectionless and connection oriented servers.

(b) How Routing sockets are different from originary sockets ? Explain.

7. Explain the following Addressing scheme :

(a) Internet adobesses.

(b) Subnet adobesses.

(c) Super net adobessing.

8. Write short notes on :

(i) Network File System

(ii) TRACERT

(iii) RPC Model

(iv) Select and Poll.

B.E.

VIth SEMESTER EXAMINATION DEC. 2008

NETWORK PROGRAMING (IT-302-E)

Time : 3 Hours *Total Marks : 100*

Note : Attempt any five questions. All question carry equal marks.

1. (a) Define computer network ? List out various benefits of computer networking ? How networking is performed in daily use ? Explain with example.

(b) Define Datagram ? How it works ? Explain format of the following datagram :

(i) IP datagram

(ii) UDP datagram

2. (a) What is ICMP ? How it works ? What is its purpose ? Explain advantages and disadvantages of ICMP ?

(b) What is Socket ? Draw structure of Socket ? Where Socket are used ? Explain with example.

3. What is Broadcasting ? Where Broadcasting is used and how ? How broadcasting is performed in mobile networks ? List out various advantages and disadvantages of Broadcasting.

4. Explain the following Socket functions with example :

(i) I/O Function

(ii) Select Function

(iii) Poll Function

How echo services are handled in TCP and UDP ? Explain.

5. What is server ? How it works ? How servers are different from TCP/IP client server ? Explain in brief various iterative and concurrent servers.

6. (a) Explain RPC with its models ? How RPC of client and server is different in terms of analogy ?

(b) Explain, how Dynamic port mapping performed in RPC.

7. What is internet ? How internet addresses are classified ? How these address are resoluted ? Explain.

8. Explain the following :

(i) Routing Sockets

(ii) RARP

(iii) RPC Transmission

(iv) Address Conversions.

❑❑❑

INDEX

INDEX

A

B

C

D

E

F

G

H

I

O

P

R

S

T

U

V

W

X

❑❑❑